SpringerWienNewYork

Günter Tiess

Legal Basics of Mineral Policy in Europe

An overview of 40 countries

SpringerWienNewYork

Dr. Günter Tiess
University Assistant at the Chair of Mining Engineering and Mineral Economics,
Montanuniversität Leoben, Austria

With financial support of
Bundesministerium für Wissenschaft und Forschung in Wien.

This work is subject to copyright.
All rights are reserved, whether the whole or part of the material is concerned, specifically those of translation, reprinting, re-use of illustrations, broadcasting, reproduction by photocopying machines or similar means, and storage in data banks.

© 2011 Springer-Verlag/Wien
Printed in Germany

SpringerWienNewYork is part of
Springer Science+Business Media
springer.at

Product Liability: The publisher can give no guarantee for all the information contained in this book. This does also refer to information about drug dosage and application there of. In every individual case the respective user must check its accuracy by consulting other pharmaceutical literature.

The use of registered names, trademarks, etc. in this publication does not imply, even in the absence of a specific statement, that such names are exempt from the relevant protective laws and regulations and therefore free for general use.

Typesetting: Jung Crossmedia Publishing GmbH, 35633 Lahnau, Germany
Printing: Strauss GmbH, 69509 Mörlenbach, Deutschland

Printed on acid-free and chlorine-free bleached paper
SPIN: 12445219

With 108 Figures and 21 Tables.

Library of Congress Control Number: 2011926822

ISBN 978-3-211-89002-8 SpringerWienNewYork

Preface

In recent years, a structural change has taken place on the global raw material markets. The old rule of thumb – 20 percent of the world population in Europe, the USA and Japan consuming more than 80 percent of the total minerals production – is not valid any more. With the integration of India, the People's Republic of China and other populous developing and emerging countries like Brazil and Russia into the world economy, today more than 50 percent of the world's population account for the largest part of mineral raw materials consumption.

Recently, the worldwide demand for mineral raw materials has increased steadily. To cope with the current global financial crisis, all countries need sufficient mineral raw materials to support attempts to boost the growth of the industrial goods production as an essential component of their reflationary programmes.

Securing the supply of raw materials has acquired a new significance for Europe. For various reasons, the public awareness of the significance of the European mineral potential has been increasing. Thus, the development of the international raw materials markets has been accompanied by the problems of mineral imports, price rises and price volatilities.

An average EU citizen's demand for mineral raw materials during his lifetime of 70 years is considerable. For example, he consumes 307 metric tonnes of sand and gravel, 39.5 t of steel, 29 t of cement, 12 t of rock salt, 1.7 t of aluminium and 1.1 t of copper.

The implementation of a coherent European minerals policy has become a tremendously important challenge which requires a consistent system of raw materials legislation.

This book places the focus on non-energy mineral raw materials.

When I decided to write this book my essential intentions were:
- to highlight the importance of the raw materials management for the EU industry,
- to address the importance of European raw materials supply security,
- to give an overview of the used (non-energy) mineral potential of Europe,
- to create an overview of the normative basics for investors.

Preface

I am also aware that the legislative basis for such an overview is subject to a process of continuous development. Therefore any suggestions from readers for amendments or corrections which they feel would improve the quality of later editions of the book would be greatly appreciated.

Finally, I would like to thank a number of people that contributed to the successful completion of this book. I am grateful to Prof. Zacharias Agioutantis at the Department of Mineral Resources Engineering, Technical University of Crete for taking the time to review the whole book and for his numerous comments and suggestions. Also, many thanks to Ms/Mr Dagestad Bård (Norway), Ben Dhonau (Ireland), Jose Antonio Espi (Spain), Tamás Hámor (Hungaria), Ken Hobden (additionally for his contribution to the introduction, United Kingdom), Paul Ike (Netherlands), Dragan Krasić (Croatia), Nadya Lazarova (Bulgaria), Martin Lutonský (Slovakia), Mihai Marinescu (Romania, Moldavia), Sokol Mati (Albania), Wolf Müller (Germany), Robert Muharremi (Kosovo), Anna Ostrega (Poland), Vladimir Simić (Serbia), Ivo Sitensky (Czech Republic), Slavko Šolar (Slovenia), Orce Spasovski (Republic of Macedonia), Caroline Strömbäck (Sweden), Cibin Ubaldo (Italy), Robert Wasserbacher (Austria) and Svetlana Zorina (Russian Federation), who contributed for each of their countries.

For assistance in completing the book, I also thank Ms. Anna Werner, Ing. Mag. Karim Karman and Ms. Irene Leischner.

Last but not least, also many thanks to my family.

Table of contents

1 Introduction 1
   1.1 The Importance of the Minerals Supply 1
   1.2 Legal Basics 3
   1.3 Requirements of a Mining Law 8

2 Albania 14
   2.1 General Facts 14
   2.2 Production of Raw Materials 14
   2.3 Normative Basics 15
       2.3.1 Primary Legal Basics 15
       2.3.2 General Rules 15
       2.3.3 Issuing of Permits 16
             2.3.3.1 Authorities 21
             2.3.3.2 Fees and Taxes 21
       2.3.4 Additional Legal Basics 22

3 Austria 24
   3.1 General Facts 24
   3.2 Production of Raw Materials 24
   3.3 Normative Basics 25
       3.3.1 Primary Legal Basics 25
             3.3.1.1 General Rules 25
             3.3.1.2 Issuing of Permits 30
             3.3.1.3 Authorities 34
             3.3.1.4 Fees and Taxes 34
       3.3.2 Additional Legal Basics 35

4 Belarus 37
   4.1 General Facts 37
   4.2 Production of Raw Materials 37
   4.3 Normative Basics 37
       4.3.1 Primary Legal Basics 37
             4.3.1.1 General Rules 38
             4.3.1.2 Issuing of Permits 38
             4.3.1.3 Authorities 39

		4.3.1.4 Fees and Taxes	40
	4.3.2	Additional Legal Basics	40

5 Belgium — 42
- 5.1 General Facts — 42
- 5.2 Production of Raw Materials — 42
- 5.3 Normative Basics — 43
  - 5.3.1 Primary Legal Basics — 43
    - 5.3.1.1 General Rules — 43
    - 5.3.1.2 Issuing of Permits — 43
    - 5.3.1.3 Authorities — 44
    - 5.3.1.4 Fees and Taxes — 44
  - 5.3.2 Additional Legal Basics — 45

6 Bulgaria — 47
- 6.1 General Facts — 47
- 6.2 Production of Raw Materials — 47
- 6.3 Normative Basics — 48
  - 6.3.1 Primary Legal Basics — 48
    - 6.3.1.1 General Rules — 48
    - 6.3.1.2 Issuing of Permits — 50
    - 6.3.1.3 Authorities — 53
    - 6.3.1.4 Fees and Taxes — 54
  - 6.3.2 Additional Legal Basics — 54

7 Croatia — 56
- 7.1 General Facts — 56
- 7.2 Production of Raw Materials — 56
- 7.3 Normative Basics — 57
  - 7.3.1 Primary Legal Basics — 57
    - 7.3.1.1 General Rules — 57
    - 7.3.1.2 Issuing of Permits — 58
    - 7.3.1.3 Authorities — 62
    - 7.3.1.4 Fees and Taxes — 62
  - 7.3.2 Additional Legal Basics — 62

8 Cyprus — 65
- 8.1 General Facts — 65
- 8.2 Production of Raw Materials — 65
- 8.3 Normative Basics — 66
  - 8.3.1 Primary Legal Basics — 66
    - 8.3.1.1 General Rules — 66
    - 8.3.1.2 Issuing of Permits — 68

8.3.1.3	Authorities	70
8.3.1.4	Fees and Taxes	70
8.3.2 Additional Legal Basics		70

9 Czech Republic 71
9.1 General Facts 71
9.2 Production of Raw Materials 71
9.3 Normative Basics 72
    9.3.1 Primary Legal Basics 72
        9.3.1.1 General Rules 72
        9.3.1.2 Issuing of Permits 77
        9.3.1.3 Authorities 79
        9.3.1.4 Fees and Taxes 79
    9.3.2 Additional Legal Basics 80

10 Denmark 82
10.1 General Facts 82
10.2 Production of Raw Materials 82
10.3 Normative Basics 83
    10.3.1 Primary Legal Basics 83
        10.3.1.1 General Rules 83
        10.3.1.2 Issuing of Permits 85
        10.3.1.3 Authorities 87
        10.3.1.4 Fees and Taxes 88
    10.3.2 Additional Legal Basics 88

11 Estonia 90
11.1 General Facts 90
11.2 Production of Raw Materials 90
11.3 Normative Basics 91
    11.3.1 Primary Legal Basics 91
        11.3.1.1 General Rules 92
        11.3.1.2 Issuing of Permits 94
        11.3.1.3 Authorities 96
        11.3.1.4 Fees and Taxes 96
    11.3.2 Additional Legal Basics 97

12 Finland 99
12.1 General Facts 99
12.2 Production of Raw Materials 99
12.3 Normative Basics 100
    12.3.1 Primary Legal Basics 100
        12.3.1.1 General Rules 100

	12.3.1.2 Issuing of Permits	102
	12.3.1.3 Authorities	105
	12.3.1.4 Fees and Taxes	105
	12.3.2 Additional Legal Basics	106

13 France 109
  13.1 General Facts 109
  13.2 Production of Raw Materials 109
  13.3 Normative Basics 110
    13.3.1 Primary Legal Basics 110
      13.3.1.1 General Rules 111
      13.3.1.2 Issuing of Permits 115
      13.3.1.3 Authorities 118
      13.3.1.4 Fees and Taxes 119
    13.3.2 Additional Legal Basics 119

14 Germany 121
  14.1 General Facts 121
  14.2 Production of Raw Materials 121
  14.3 Normative Basics 123
    14.3.1 Primary Legal Basics 123
      14.3.1.1 General Rules 123
      14.3.1.2 Issuing of Permits 129
      14.3.1.3 Authorities 131
      14.3.1.4 Fees and Taxes 132
    14.3.2 Additional Legal Basics 132

15 Greece 134
  15.1 General Facts 134
  15.2 Production of Raw Materials 134
  15.3 Normative Basics 135
    15.3.1 Primary Legal Basics 135
      15.3.1.1 General Rules 136
      15.3.1.2 Issuing of Permits 137
      15.3.1.3 Authorities 138
      15.3.1.4 Fees and Taxes 139
    15.3.2 Additional Legal Basics 140

16 Hungary 142
  16.1 General Facts 142
  16.2 Production of Raw Materials 142
  16.3 Normative Basics 143
    16.3.1 Primary Legal Basics 143

16.3.1.1 General Rules	143
16.3.1.2 Issuing of Permits	145
16.3.1.3 Authorities	148
16.3.1.4 Fees and Taxes	149
16.3.2 Additional Legal Basics	150

17 Iceland 152
17.1 General Facts 152
17.2 Production of Raw Materials 152
17.3 Normative Basics 153
    17.3.1 Primary Legal Basics 153
        17.3.1.1 General Rules 153
        17.3.1.2 Issuing of Permits 153
        17.3.1.3 Authorities 154
        17.3.1.4 Fees and Taxes 155
    17.3.2 Additional Legal Basics 155

18 Ireland 156
18.1 General Facts 156
18.2 Production of Raw Materials 156
18.3 Normative Basics 157
    18.3.1 Primary Legal Basics 157
        18.3.1.1 General Rules 157
        18.3.1.2 Issuing of Permits 160
        18.3.1.3 Authorities 163
        18.3.1.4 Fees and Taxes 164
    18.3.2 Additional Legal Basics 164

19 Italy 165
19.1 General Facts 165
19.2 Production of Raw Materials 165
19.3 Normative Basics 166
    19.3.1 Primary Legal Basics 166
        19.3.1.1 General Rules 166
        19.3.1.2 Issuing of Permits 167
        19.3.1.3 Authorities 171
        19.3.1.4 Fees and Taxes 171
    19.3.2 Additional Legal Basics 172

20 Kosovo 177
20.1 General Facts 177
20.2 Production of Raw Materials 177
20.3 Normative Basics 178

20.3.1 Primary Legal Basics		178
20.3.1.1 General Rules		178
20.3.1.2 Issuing of Permits		183
20.3.1.3 Authorities		187
20.3.1.4 Fees and Taxes		187
20.3.2 Additional Legal Basics		188

21 Latvia … 190
- 21.1 General Facts … 190
- 21.2 Production of Raw Materials … 190
- 21.3 Normative Basics … 191
  - 21.3.1 Primary Legal Basics … 191
    - 21.3.1.1 General Rules … 191
    - 21.3.1.2 Issuing of Permits … 192
    - 21.3.1.3 Authorities … 194
    - 21.3.1.4 Fees and Taxes … 194
  - 21.3.2 Additional Legal Basics … 194

22 Lithuania … 196
- 22.1 General Facts … 196
- 22.2 Production of Raw Materials … 196
- 22.3 Normative Basics … 197
  - 22.3.1 Primary Legal Basics … 197
    - 22.3.1.1 General Rules … 197
    - 22.3.1.2 Issuing of Permits … 199
    - 22.3.1.3 Authorities … 202
    - 22.3.1.4 Fees and Taxes … 202
  - 22.3.2 Additional Legal Basics … 203

23 Luxembourg … 204
- 23.1 General Facts … 204
- 23.2 Normative Basics … 204
  - 23.2.1 Primary Legal Basics … 204
    - 23.2.1.1 General Rules … 204
    - 23.2.1.2 Issuing of Permits … 205
    - 23.2.1.3 Authorities … 205
    - 23.2.1.4 Fees and Taxes … 206
  - 23.2.2 Additional Legal Basics … 206

24 Macedonia (Former Yugoslav Republic of Macedonia – FYROM) … 208
- 24.1 General Facts … 208
- 24.2 Production of Raw Materials … 208
- 24.3 Normative Basics … 209

24.3.1 Primary Legal Basics	209
24.3.1.1 General Rules	209
24.3.1.2 Issuing of Permits	217
24.3.1.3 Authorities	220
24.3.1.4 Fees and Taxes	220
24.3.2 Additional Legal Basics	221

25 Malta — 222
- 25.1 General Facts — 222
- 25.2 Production of Raw Materials — 222
- 25.3 Normative Basics — 222
  - 25.3.1 Primary Legal Basics — 222
    - 25.3.1.1 General Rules — 223
    - 25.3.1.2 Issuing of Permits — 226
    - 25.3.1.3 Authorities — 227
    - 25.3.1.4 Fees and Taxes — 227
  - 25.3.2 Additional Legal Basics — 228

26 Moldavia — 229
- 26.1 General Facts — 229
- 26.2 Production of Raw Materials — 229
- 26.3 Normative Basics — 230
  - 26.3.1 Primary Legal Basics — 230
    - 26.3.1.1 General Rules — 230
    - 26.3.1.2 Issuing of Permits — 234
    - 26.3.1.3 Authorities — 235
    - 26.3.1.4 Fees and Taxes — 235
  - 26.3.2 Additional Legal Basics — 235

27 Netherlands — 237
- 27.1 General Facts — 237
- 27.2 Production of Raw Materials — 237
- 27.3 Normative Basics — 238
  - 27.3.1 Primary Legal Basics — 238
    - 27.3.1.1 General Rules — 238
    - 27.3.1.2 Issuing of Permits — 244
    - 27.3.1.3 Authorities — 246
    - 27.3.1.4 Fees and Taxes — 247
  - 27.3.2 Additional Legal Basics — 247

28 Norway — 248
- 28.1 General Facts — 248
- 28.2 Production of Raw Materials — 248

  28.3 Normative Basics 249
    28.3.1 Primary Legal Basics 249
      28.3.1.1 Issuing of Permits 252
      28.3.1.2 Authorities 257
      28.3.1.3 Fees and Taxes 258
    28.3.2 Additional Legal Basics 258

29 Poland 260
  29.1 General Facts 260
  29.2 Production of Raw Materials 260
  29.3 Normative Basics 261
    29.3.1 Primary Legal Basics 261
      29.3.1.1 General Rules 262
      29.3.1.2 Issuing of Permits 264
      29.3.1.3 Authorities 265
      29.3.1.4 Fees and Taxes 266
    29.3.2 Additional Legal Basics 267

30 Portugal 269
  30.1 General Facts 269
  30.2 Production of Raw Materials 269
  30.3 Normative Basics 270
    30.3.1 Primary Legal Basics 270
      30.3.1.1 General Rules 270
      30.3.1.2 Issuing of Permits 276
      30.3.1.3 Authorities 279
      30.3.1.4 Fees and Taxes 279
    30.3.2 Additional Legal Basics 281

31 Romania 282
  31.1 General Facts 282
  31.2 Production of Raw Materials 282
  31.3 Normative Basics 283
    31.3.1 Primary Legal Basics 283
      31.3.1.1 General Rules 283
      31.3.1.2 Issuing of Permits 284
      31.3.1.3 Authorities 287
      31.3.1.4 Fees and Taxes 287
    31.3.2 Additional Legal Basics 288

32 Russian Federation 290
  32.1 General Facts 290
  32.2 Production of Raw Materials 290

32.2.1 Primary Legal Basics	291
32.2.1.1 General Rules	291
32.2.1.2 Issuing of Permits	294
32.2.1.3 Authorities	296
32.2.1.4 Fees and Taxes	297
32.2.2 Additional Legal Basics	299

33 Serbia — 301

- 33.1 General Facts — 301
- 33.2 Production of Raw Materials — 301
- 33.3 Normative Basics — 302
  - 33.3.1 Primary Legal Basics — 302
    - 33.3.1.1 General Rules — 303
    - 33.3.1.2 Issuing of Permits — 304
    - 33.3.1.3 Authorities — 307
    - 33.3.1.4 Fees and Taxes — 308
  - 33.3.2 Additional Legal Basics — 308

34 Slovakia — 310

- 34.1 General Facts — 310
- 34.2 Production of Raw Materials — 310
- 34.3 Normative Basics — 311
  - 34.3.1 Primary Legal Basics — 311
    - 34.3.1.1 General Rules — 311
    - 34.3.1.2 Issuing of Permits — 315
    - 34.3.1.3 Authorities — 316
    - 34.3.1.4 Fees and Taxes — 316
  - 34.3.2 Additional Legal Basics — 317

35 Slovenia — 318

- 35.1 General Facts — 318
- 35.2 Production of Raw Materials — 318
- 35.3 Normative Basics — 319
  - 35.3.1 Primary Legal Basics — 319
    - 35.3.1.1 General Rules — 319
    - 35.3.1.2 Issuing of Permits — 327
    - 35.3.1.3 Authorities — 329
    - 35.3.1.4 Fees and Taxes — 330
  - 35.3.2 Additional Legal Basics — 330

36 Spain — 333

- 36.1 General Facts — 333
- 36.2 Production of Raw Materials — 333

Table of contents

36.3 Normative Basics	334
36.3.1 Primary Legal Basics	334
36.3.1.1 General Rules	335
36.3.1.2 Issuing of Permits	339
36.3.1.3 Authorities	341
36.3.1.4 Fees and Taxes	342
36.3.2 Additional Legal Basics	343
37 Sweden	**344**
37.1 General Facts	344
37.2 Production of Raw Materials	344
37.3 Normative Basics	345
37.3.1 Primary Legal Basics	345
37.3.1.1 General Rules	345
37.3.1.2 Issuing of Permits	348
37.3.1.3 Authorities	353
37.3.1.4 Fees and Taxes	354
37.3.2 Additional Legal Basics	354
38 Switzerland	**356**
38.1 General Facts	356
38.2 Production of Raw Materials	356
38.3 Normative Basics	357
38.3.1 Primary Legal Basics	357
38.3.1.1 General Rules	357
38.3.1.2 Issuing of Permits	358
38.3.1.3 Authorities	360
38.3.1.4 Fees and Taxes	360
38.3.2 Additional Legal Basics	360
39 Turkey	**362**
39.1 General Facts	362
39.2 Production of Raw Materials	362
39.3 Normative Basics	364
39.3.1 Primary Legal Basics	364
39.3.1.1 General Rules	364
39.3.1.2 Issuing of Permits	367
39.3.1.3 Authorities	369
39.3.1.4 Fees and Taxes	369
39.3.2 Additional Legal Basics	370

40 Ukraine ... 372
- 40.1 General Facts ... 372
- 40.2 Production of Raw Materials ... 372
- 40.3 Normative Basics ... 373
  - 40.3.1 Primary Legal Basics ... 373
    - 40.3.1.1 General Rules ... 373
    - 40.3.1.2 Issuing of Permits ... 375
    - 40.3.1.3 Authorities ... 375
    - 40.3.1.4 Fees and Taxes ... 376
  - 40.3.2 Additional Legal Basics ... 376

41 United Kingdom ... 378
- 41.1 General Facts ... 378
- 41.2 Production of Raw Materials ... 378
- 41.3 Normative Basics ... 379
  - 41.3.1 Primary Legal Basics ... 379
    - 41.3.1.1 General Rules ... 379
    - 41.3.1.2 Issuing of Permits ... 380
    - 41.3.1.3 Authorities ... 383
    - 41.3.1.4 Fees and Taxes ... 384
  - 41.3.2 Additional Legal Basics ... 384

42 References ... 387

43 Additional updated data directly from the concerned countries ... 390

List of figures and tables

Figure 1: Minerals demand of an EU citizen. (BGR) 1
Figure 2: Laws concerning mineral extraction – Germany 126
Figure 3: Overview of approval procedures – Germany (Müller, Schulz, 2000).. 127

Table 1: Important producers of metallic raw materials in the EU, including Turkey (European Commission, 2007) 3
Table 2: Mineral resource legislation in selected countries of Europe 4
Table 3: Production Data – Albania (World Mining Data, Weber and Zsak, 2008)... 14
Table 4: Structure of the Mining Law of Albania 15
Table 5: List of Acts significant for raw materials – Albania. 22
Table 6: Production Data– Austria (World Mining Data, L. Weber and Zsak, 2008)... 24
Table 7: Aggregates Production – Annual Statistics/Austria, 22 April 2008 (UEPG 2008)....................................... 25
Table 8: Structure of the Mining Law ("Mineralrohstoffgesetz") – Austria 25
Table 9: Mineral Rights according to the Mining Law................. 29
Table 10: Fees according to the mineral rights (for each calendar year) ... 34
Table 11: List of Acts significant for Raw Materials – Austria 35
Table 12: Production Data – Belarus (World Mining Data, Weber and Zsak, 2008)... 37
Table 13: Production of lime – Belarus (USGS) 37
Table 14: Structure of the Mining Law – Belarus 38
Table 15: List of Acts significant for raw materials – Belarus 40
Table 16: Production Data – Belgium (World Mining Data, Weber and Zsak)... 42
Table 17: Aggregates Production – Annual Statistics/Belgium, 22 April 2008 (UEPG 2008)....................................... 42
Table 18: List of Acts significant for raw materials – Belgium 45
Table 19: Production Data – Bulgaria (World Mining Data, Weber and Zsak, 2008)... 47
Table 20: Production of aggregates – Bulgaria (USGS, 2008) 48
Table 21: Structure of the Mining Law 1999 – Bulgaria 48

Table 22: List of acts significant for raw materials – Bulgaria 54
Table 23: Production Data – Croatia (World Mining Data, Weber and Zsak, 2008). 56
Table 24: Aggregates Production – Annual Statistics/Croatia, 22 April 2008 (UEPG 2008) 56
Table 25: Structure of the Mining Law – Croatia 57
Table 26: List of acts significant for raw materials – Croatia............ 62
Table 27: Production Data – Cyprus (World Mining Data, Weber and Zsak, 2008). 65
Table 28: Production of aggregates – Cyprus (USGS, 2008)............ 65
Table 29: Structure of the Mining Law – Cyprus 66
Table 30: List of Acts significant for raw materials – Cyprus 70
Table 31: Production Data – Czech Republic (World Mining Data, Weber and Zsak, 2008). 71
Table 32: Aggregates Production – Annual Statistics/Czech Republic, 22. April 2008 (UEPG 2008) 72
Table 33: Structure of the Mining Law – Czech Republic 72
Table 34: List of Acts significant for raw materials – Czech Republic 80
Table 35: Production Data – Denmark (World Mining Data, Weber and Zsak, 2008). 82
Table 36: Aggregates Production – Annual Statistics/Denmark, 22 April 2008 (UEPG 2008) 82
Table 37: Structure of the Mining Law – Denmark 83
Table 38: List of Acts significant for raw materials – Denmark 88
Table 39: Production Data – Estonia (World Mining Data, Weber and Zsak, 2008). 90
Table 40: Production of aggregates – Estonia (USGS, 2008)............ 90
Table 41: Structure of the Mining Law 2003, Estonia 92
Table 42: List of Acts significant for raw materials – Estonia 97
Table 43: Production Data – Finland (World Mining Data, Weber and Zsak, 2008). 99
Table 44: Aggregates Production – Annual Statistics/Finland, 22 April 2008 (UEPG, 2008) 100
Table 45: Structure of the Mining Act of Finland..................... 100
Table 46: List of Acts significant for raw materials – Finland........... 106
Table 47: Production Data – France (World Mining Data, Weber and Zsak, 2008). 109
Table 48: Aggregates Production – Annual Statistics/France, 22 April 2008 (UEPG 2008) 110
Table 49: Structure of the Code Minier, France 111
Table 50: List of Acts significant for raw materials – France............ 119

List of figures and tables

Table 51: Production Data – Germany (World Mining Data, Weber and Zsak, 2008). ... 121
Table 52: Aggregates Production – Annual Statistics/Germany, 22 April 2008 (UEPG 2008) ... 122
Table 53: Structure of the Mining Law – Germany ... 123
Table 54: Mining Licenses for free for mining minerals – Germany (Müller, Schulz 2000) ... 129
Table 55: List of acts significant for raw materials – Germany ... 132
Table 56: Production Data – Greece (World Mining Data, Weber and Zsak, 2008). ... 134
Table 57: Production of aggregates – Greece (USGS, 2008). ... 135
Table 58: Responsibility for issuing mineral permissions ... 138
Table 59: List of acts significant for raw materials – Greece ... 140
Table 60: Production Data – Hungary (World Mining Data, Weber and Zsak, 2008). ... 142
Table 61: Production of aggregates – Hungary (USGS, 2008). ... 143
Table 62: Structure of the Mining Law – Hungary. ... 143
Table 63: List of Acts significant for raw materials – Hungary. ... 150
Table 64: Production Data – Iceland (World Mining Data, Weber and Zsak). ... 152
Table 65: Production of aggregates – Iceland (USGS, 2008) ... 152
Table 66: List of Acts significant for raw materials – Iceland ... 155
Table 67: Production Data – Ireland (World Mining Data, Weber and Zsak, 2008). ... 156
Table 68: Aggregates Production – Annual statistics/Ireland, 22 April 2008 (UEPG 2008) ... 156
Table 69: Structure of the Irish Mining Law ("Minerals Development Act"). ... 157
Table 70: List of Acts significant for raw materials – Irland ... 164
Table 71: Production Data – Italy (World Mining Data, Weber and Zsak, 2008). ... 165
Table 72: Aggregates Production – Annual Statistics/Italy, 22 April 2008 (UEPG 2008). ... 166
Table 73: Structure of the Regio decreto 29 luglio 1927, n. 1443 ... 166
Table 74: List of Acts significant for raw materials – Italy (Badino, G. et al, 2004). ... 172
Table 74a: Regional laws on quarrying activities ... 175
Table 75: Production data – Kosovo (USGS). ... 177
Table 76: Structure of the Mining Regulation (MR) – Kosovo. ... 178
Table 77: List of Acts significant for Raw Materials – Kosovo ... 188

List of figures and tables

Table 78: Production Data – Latvia (World Mining Data, Weber and Zsak, 2008). 190
Table 79: Production of aggregates – Latvia (USGS, 2008) 190
Table 80: Structure of the Mining Law – Latvia 191
Table 81: List of Acts significant for raw materials – Latvia 194
Table 82: Production Data – Lithuania (World Mining Data, Weber and Zsak, 2008). 196
Table 83: Production of aggregates – Lithuania (USGS) 196
Table 84: Structure of the Mining Law – Lituania 197
Table 85: List of Acts significant for raw materials – Lithuania 203
Table 86: List of Acts significant for raw materials – Luxembourg 206
Table 87: Production Data – Macedonia (World Mining Data, Weber and Zsak, 2008). 208
Table 88: Production of bentonite, Gypsum, lime and construction minerals – Macedonia (USGS, 2008). 209
Table 89: Structure of the Mining Law – Macedonia 209
Table 90: List of Acts significant for raw materials – Macedonia 221
Table 91: Production Data – Malta (World Mining Data, Weber and Zsak, 2008). 222
Table 92: Production of limestone – Malta (USGS, 2008) 222
Table 93: Structure of the Malta Resources Authority Act XXV of 2000 223
Table 94: List of Acts significant for raw materials – Malta. 228
Table 95: Production Data – Moldavia (World Mining Data, Weber and Zsak, 2008). 229
Table 96: Production of lime and gravel – Moldavia (USGS, 2008) 229
Table 97: Structure of the Mining Law – Moldavia. 230
Table 98: List of Acts significant for raw materials – Moldavia. 235
Table 99: Production Data – Netherlands (World Mining Data, Weber and Zsak, 2008) 237
Table 100: Aggregates Production – Annual Statistics/Netherlands, 22. April 2008 (UEPG 2008) 237
Table 101: Structure of the Mining Law (2003). 238
Table 102: List of Acts significant for raw materials – Netherlands. 247
Table 103: Production Data – Norway (World Mining Data, Weber and Zsak, 2008). 248
Table 104: Aggregates Production – Annual Statistics/Norway, 22 April 2008 (UEPG 2008) 249
Table 105: Structure of the Mining Law – Norway 249
Table 106: List of Acts significant for Raw Materials – Norway 258
Table 107: Production Data – Poland (World Mining Data, Weber and Zsak, 2008). 260

List of figures and tables

Table 108: Aggregates Production – Annual Statistics/Poland, 22 April 2008 (UEPG 2008) 261
Table 109: Structure of the Geological and Mining Law – Poland 262
Table 110: List of Acts significant for Raw Materials – Poland 267
Table 111: Production Data – Portugal (World Mining Data, Weber and Zsak, 2008). 269
Table 112: Aggregates Production – Annual Statistics/Portugal, 22 April 2008 (UEPG 2008) 270
Table 113: Structure of the Mining Law of 1990. 270
Table 114: Taxes in accordance with Article 66 of Decree No. 88 of 1990 (Magno, 2004) 279
Table 115: Articles, Assignments and Taxes according (Article 56) Law No. 89 of 1990 (Magno, 2004) 280
Table 116: List of Acts significant for raw materials – Portugal (Magno, 2004). 281
Table 117: Production Data – Romania (World Mining Data, Weber and Zsak, 2008). 282
Table 118: Aggregates Production – Annual Statistics/Romania, 22. April 2008 (UEPG 2008) 283
Table 119: Structure of the Mining Law – Romania. 283
Table 120: List of Acts significant for Raw Materials – Romania. 288
Table 121: Production Data – Russian Federation (European Section) (World Mining Data, Weber and Zsak, 2008) 290
Table 122: Production of lime – Russian Federation (USGS, 2008) 291
Table 123: Structure of the Law of the Russian Federation on Subsoil ... 291
Table 124: Rates of Regular Payments for the Use of Subsoil (Article 43 ML) – Russian Federation 298
Table 125: List of Acts significant for Raw Materials – Russian Federation 299
Table 126: Production Date – Serbia (World Mining Data, Weber and Zsak, 2008). 301
Table 127: Production of aggregates – Serbia (USGS, 2008) 302
Table 128: Structure of Law on Geological Exploration – Serbia 303
Table 129: Structure of the Mining Law – Serbia 303
Table 130: List of Acts significant for Raw Materials – Serbia 308
Table 131: Production Data – Slovakia (World Mining Data, Weber and Zsak, 2008). 310
Table 132: Aggregates Production – Annual Statistics/Slovakia, 22 April 2008 (UEPG 2008) 311
Table 133: Structure of the Mining Law 311
Table 134: List of Acts significant for raw materials – Slovakia.......... 317

List of figures and tables

Table 135: Production Data – Slovenia (World Mining Data, Weber and Zsak, 2008)...	318
Table 136: Production of aggregates – Slovenia (Metric tons unless otherwise specified), (Source: USGS, 2009).....................	318
Table 137: Structure of the Mining Law – Slovenia	319
Table 138: List of Acts significant for raw materials – Slovenia	330
Table 139: Production Data – Spain (World Mining Data, Weber and Zsak, 2008)...	333
Table 140: Aggregates Production – Annual Statistics/Spain, 22 April 2008, quantities in million tonnes, (Source: UEPG 2008)............	334
Table 141: Structure of the Mining Law – Spain	335
Table 142: Taxes according Article 66 of Decree No. 88 of 1990	342
Table 143: Taxes according to Article 56 of Decree No. 89 of 1990 (construction minerals)	342
Table 144: List of Acts significant for raw materials – Spain (source: Espi, 2004)...	343
Table 145: Production Date – Sweden (World Mining Data, Weber and Zsak, 2008)...	344
Table 146: Aggregates Production – Annual Statistics/Sweden, 22 April 2008 (UEPG 2008)	345
Table 147: Structure of the Mining Law – Sweden	345
Table 148: List of Acts significant for Raw Materials – Sweden	354
Table 149: Production Data – Switzerland (World Mining Data, Weber and Zsak, 2008)...	356
Table 150: Aggregates Production – Annual Statistics/Switzerland, 22. April 2008 (UEPG 2008)	356
Table 151: List of acts significant for raw materials – Switzerland	352
Table 152: Legal basics – Swiss construction minerals industry (Grob, 2005)..	360
Table 153: Production Data – Turkey (World Mining Data, Weber and Zsak, 2008)	362
Table 154: Aggregates Production – Annual Statistics/Turkey, 22 April 2008 (UEPG 2008)	363
Table 155: Structure of the Mining Law of Turkey	364
Table 156: List of Acts significant for raw materials – Turkey	370
Table 157: Production Data – Ukraine (World Mining Data, Weber and Zsak, 2008)	372
Table 158: Production of clays and gypsum – Ukraine (USGS, 2008)....	373
Table 159: Structure of the Mining Law – Ukraine	373
Table 160: Law No. 1127 of 1999 – Ukraine	374
Table 161: List of Acts significant for Raw Materials – Ukraine.........	376

List of figures and tables

Table 162: Production Data – UK (World Mining Data, Weber and Zsak, 2008) 378
Table 163: Aggregates Production – Annual Statistics/United Kingdom, 22. April 2008 (UEPG 2008) 379
Table 164: List of Acts significant for raw materials – UK (Ike, 2004) ... 384

Abbreviations used in this book

cp.	compare
e.g.	for example
et al.	and others
ibidem	in the same place
lc.	at the specified location

1 Introduction

1.1 The Importance of the Minerals Supply

Raw materials are needed for many purposes. For the making of a computer more than 30 **metallic and non-metallic** raw materials are needed. The production of a car requires more than 40 different raw materials, such as aluminium, iron and zinc for the car body, platinum or palladium for the catalyst and copper for the on-board electronics. Even a mobile phone contains 40 different raw materials, for example lithium, tantalum, cobalt and antimony.

Likewise, the **industrial minerals** are of wide-ranging significance. They are indispensable for most industrial processes. Calcium carbonate can be used as a filler in paper manufacture and plastics processing, as an extender in paints and varnishes, as a fertilizer in agriculture and for the desulphurisation of waste gases as an environmental measure. It is also used as an additive in the glass and ceramics industries and in cleaning and cosmetic products. Calcium carbonate is even to be found in toothpaste.

As **Construction raw materials,** minerals are used in a wide range of applications, for road, track and sewer construction as well as for residential, office and industrial building. The construction of a one-family house with basement requires approximately 450 tonnes of raw materials. An apartment of 80m² uses 100 tonnes of raw materials. For a single kilometre of highway about 160,000 tonnes of raw materials are needed.

Figure 1: Mineral demand of an EU Citizen – consumption or use of mineral raw materials and energy resources in Germany over a lifetime/age 89 years, based of 2008 data (BGR)

Consumption of an average European citizen

SAND AND GRAVEL	307 t	SILICA SAND	4.7 t
BROWN COAL	158 t	KAOLIN	4.0 t
CRUSHED ROCK	130 t	POTASH (K_2O)	3.4 t
PETROLEUM	116 t	ALUMINIUM	1.7 t
NATURAL GAS (1000 m³)	89.6	COPPER	1.1 t

1 Introduction

LIMESTONE, DOLOMITE	72 t	STEEL ALLOYING METALS	0.9 t
HARD COAL	67 t	SULPHUR	0.2 t
CRUDE STEEL	39.5 t	ASBESTOS	0.16 t
CEMENT	29 t	PHOSPHATE ROCK	0.15 t
ROCK SALT	12 t	ELECTRICITY (MWh)	293.2
GYPSUM	8.5		

Significance of mineral policy

For various reasons, mineral policy has become a major topic in Europe.[1] Movements of the international commodity markets are prominent indicators. These movements are influenced by the same factors which also in the past repeatedly caused disruptions to the markets. However, since the beginning of this decade, these factors have had occurred with intensified influence and hence greater leverage. Without doubt, the dissolution of the Soviet Union and the consequent opening of the international trade markets for mineral commodities at first gave rise to a period of good availability and thus lower prices.[2] Not unexpectedly, new consumers with lively demand for raw materials have emerged on the markets. It has been predicted for a long time that particularly the countries of the Circum-Pacific area would experience a strong economic growth. It was to be foreseen that these countries would enter the global market and would not only influence as low-cost suppliers of consumer goods, but with the subsequent increase in prosperity their inhabitants would also make higher demands concerning their own living conditions and infrastructure.[3]

In recent years, the **global demand for raw materials has been increasing continuously and presently is at the beginning of a new growth curve.** In order to cope with the global financial crisis, several countries have attempted to push the growth of industrial goods production in line with economic stimulus packages. It is logical that the demand for raw materials must remain at a high level as a consequence. Against the background of these developments in the international commodity markets, it is important to note the mineral potential (and investing potential) of the European nations.

---

1 Cf. The Raw Materials Initiative – meeting our critical needs for growth and jobs in Europe.
2 Vereinigung Rohstoffe und Bergbau (VRB), (2008): 2008, Positionen und Perspektiven.
3 lc.

Table 1: Important producers of metallic raw materials in the EU, including Turkey (Source: European Commission, 2007)

Metal	% worldwide	Countries with > 1% of World production in 2005
Silver	9,5	Poland (6,2%), Sweden, Turkey
Zinc	9,4	Ireland (4,4%), Sweden, Poland
Titanium	7,4	Norway (7,4%)
Lead	6,8	Ireland (2,2%), Sweden, Poland
Chromium	6,7	Finland (3,6%), Turkey
Copper	5,6	Poland (3,4%)
Tungsten	3,8	Austria (2,3%), Portugal
Nickel	2,1	Greece (1,4%)
Iron	2,0	Sweden (1,5%)
Aluminium (Bauxite)	2,0	Greece (1,4%)
Mercury	1,8	Finland (1,8%)

The **production data on metallic raw materials and industrial minerals** of the countries featured in this book are based on the World Mining Data by Weber and Zsak (2008). The production data on construction raw materials are also listed for the respective countries where available.[4] Annual production of construction raw materials (more than 3 billion tonnes) is of particular importance for Europe.

1.2 Legal Basics

A consistent body of rules is in general of central importance to a secure supply of raw materials. On one hand, it provides investment and legal security for entrepreneurs. On the other hand, it gives regulatory authorities clear guidelines for the enforcement of laws and observance of safety and environmental standards.

---

[4] Different sources used: Union Europeenne des Producteurs de Granulats, UEPG (2008), Construction raw materials production 2006 – Annual statistics, April 22, 2008. Where no data is available from UEPG, information from USGS (2008): Minerals Yearbook, is used.

1 Introduction

The legal basis for minerals consists of the primary legal basis and other legal bases relevant to the matter of raw materials. This can be illustrated by the following table (Table 2) of the various laws in place across Europe.[5]

The table concentrates on the primary legal basis but refers only to the respective laws on mineral resources applying in each country. All raw materials are not necessarily included in a uniform mineral resources law. It is not unusual to decouple specific raw materials from the general mineral regulations. Within the body of mineral rules, it should be made clear precisely which raw materials are covered and which are not.[6]

Table 2: Mineral resource legislation in selected countries of Europe

Countries	Main basic laws related to mining Laws	Versions	Remarks
Albania	Mining Law No. 7796 of 1994, as amended by Law 6291 of 2004	English	
Austria	Mining Law No. 38 of 1999 as amended by Law.115 of 2009	German	
Belarus	Subsoil Code Law No. 103-Z of 1999	Russian	
Belgium			No Mining Law; environmental laws are the main legal basic
Bulgaria	Law on Mineral Resources No. 23 of 1999	English	
Croatia	Mining Law No. 35 of 1995	English	
Cyprus	Mining Law ("Mines and Quarries Law") No. 5 of 1965, as amended by Law No. 63 of 2003	English	
Czech Republic	Mining Act No. 44 of 1988, as amended by Law No. 186 of 2006	English	
Denmark	("Raw Materials Act") No. 569 of 1997, as amended by Law No. 145 of 2002	English	

---

5   There is no claim to completeness.
6   E. g. Norway.

1.2 Legal Basics

Countries	Main basic laws related to mining Laws	Versions	Remarks
Estonia	Mining Law No. 20, 118 of 2003 amended by the Law No. 18, 131 of 2004	English	Earth's Crust Act, relevant for quarries
Finland	Mining Law No. 1965 of 503 as amended in 1997	English (unofficial translation)	Land Extraction Act No. 555 of 1981, relevant for quarries.
France	Mining Code ("Code Minier"; 1992, amended by Law 2004/105 and regulation 2006/407	French	Law 1993/93-3, Loi sur les Installations Classees, relevant for quarries
Germany	Mining Law No. 1310 of 1980, as amended by Law No. 2585 of 2009	German	
Greece	Decree No. 210 of 1973 commonly referred to as "The Mining Code" as amended by Law No. 274 of 1976	Greek	No single mining law, All provisions concerning the exploitation of industrial minerals and aggregates are defined by the Regulations on Mining and Quarrying Activities
Hungary	Mining Law No. XLVIII of 1993 as amended by Law No. CXXXIII of 2007	English	
Iceland	Nature Conservation Law No. 44 of 1999	English	
Ireland	Minerals Development Act No. 31 of 1940 as amended by Law No. 21 of 1999	English	
Italy	Mining Law (Regio Decreto) No. 1443 of 1927 as amended by Legislative Decree No. 213 of 1999	Italian	Separate quarrying laws of regions
Kosovo	Law No. 03/L-163 on Mines and Minerals of 2010	English	

5

1 Introduction

Countries	Main basic laws related to mining Laws	Versions	Remarks
Latvia	Law on the Subsoil No. 13 of 1996 as amended by Law No. 321/322 of 2000	English (unofficial translation)	
Lithuania	Underground Law No. VIII-573 of 1995	English	
Luxembourg			No Mining Law; environmental laws are the main legal basic
Macedonia	Law on Minerals Raw Materials	English	
Malta	Malta Resources Authority Act XXV of 2000, as amended by Act XXIII of 2009	English	
Moldova	Subsoil Code No. 1511-XII of 1993		
Netherlands	Mining Law (Mining Act) 2003 and Excavation Act 2002 (regulating industrial and construction minerals)	English (unofficial translation of the Mining Act)	
Norway	Mining Act No. 101 of 2009	English	Planning and Building Act (construction minerals) No. 77 of 1985
Poland	Law on Geology and Mining No. 27 of 1994 as amended by Law No. 138, 154, 199 of 2008	English	
Portugal	Mining Law No. 18 713 of 1930 as amended by Law No. 90 of 1990	Portuguese	Ore deposits Law (Act No. 88/90) and Quarry Law (Act No. 89/90 modified on 2001)
Romania	Mining Law No. 85 of 2003	English (unofficial translation)	

1.2 Legal Basics

Countries	Main basic laws related to mining Laws	Versions	Remarks
Russian Federation	Law on Subsoil No. 2395 of 1992 as amended in 2007	English	
Slovakia	Mining Act No. 44 of 1988 as amended by Law 498 of 1991	English	
Slovenia	Mining Act No. 56 of 1999 as amended by Law 68 of 2008	English	
Spain	Mining Law No. 22 of 1973 as amended by Law No. 54 of 1980	Spanish	
Sweden	Mining Law No. 45 of 1991 as amended by Law No. 943 of 2005	English (unofficial translation)	
Switzerland	Mining Law, Kanton Bern, 1979	German	
Serbia	Mining Law No. 5 of 2006	English (unofficial translation)	Law on Geological Exploration No. 44 of 1995
Turkey	Mining Law No. 3213 of 1985, as amended by Law No. 5177 of 2004	English	
Ukraine	Subsoil Code No. 132 of 1994	Russian	
UK			Different laws are relevant

A quick overview of the systems used in individual countries, the structure of the respective law on mineral resources (i. e. mining law) is presented in a schematic manner. The specific laws on mineral resources are discussed according to particular criteria which are equally fixed for each country as follows:[7]

- General regulations
  - Mineral groups and mining rights
- Issuing of permits
  - Prospection, exploration, exploitation
- Authorities
- Taxes and fees

---

[7] This concerns only selected issues, i. e not all regulations of a specific law are discussed.

1.3 Requirements of a Mining Law

Modern mineral resource law has to perform several functions. The business of providing mineral raw materials is complex, in particular in relation to increasingly higher standards of environmental protection. Key aspects, standards and principles which must be addressed by that law are considered in the following paragraphs.[8]

New mining projects generally pass through various stages. Using the example of aluminium; the first phase is prospection (search) for a deposit, followed by **exploration** (investigating and/or evaluation), **exploitation** (of bauxite) and mineral **processing** (to produce aluminium concentrate). The final product is aluminium. For the regulation of such a complex development process a solid legal basis is necessary, which is of great significance both to the investor (entrepreneur) and to the authorities.

Mineral extraction is characterised by its location binding, forecast uncertainty, the dynamic mode of operation, its temporary use of the land surface, the sequence of prospecting, exploration, operating and closure phases as well as some risks and dangers that are specific to mining. To deal with these characteristics, specific regulations as contained in a mineral resource law are required.

To provide a secure supply of raw materials, particularly in the face of today's rising market prices, demands that a balance is struck between the different interests by means of the regulatory system. Modern mineral resource law, together with any accompanying regulations, takes into account the principle of sustainability in which economic, social and environmental concerns are considered in the context of an activity that is location bound.

Therefore an essential part of the regulation of mining activities is not only the extraction process but also the phase of the closure of mines; the interests of adequate restoration after the termination of exploitation. Operational responsibility only comes to an end when acceptable conditions for reuse of the mining area are created and no dangers created by mining remain.

---

8   Otto, J. M. (1999): Mining, Environment and Development, United Nations Conference on Trade and Development, USA. Mihatsch, A. (2002): Mineralrohstoffgesetz (MinroG). 2., rev. ed. Manz, Wien, 2002. Kullmann, U., Requirements for a modern mining law in the framework of European legislation, Federal Ministry of Economics and Technology, Germany. Müller, W and Schulz, P.-M. (2000): Handbuch Recht der Bodenschätzegewinnung, Baden-Baden, Deutschland. Vereinigung Rohstoffe und Bergbau (VRB), (2008): Positionen und Perspektiven, 2008.

1.3 Requirements of a Mining Law

Mining Law and other legal bases

The raw materials sector is regulated by a range of different laws and authorities, necessitating a similarly wide range of interaction. A mining operation might be covered by mining law, environmental law, spatial planning law, employment, commercial and financial law. Mineral resources law should, considering the potential for conflict, provide a comprehensive code of regulations for the various points of contact in all regulatory fields: this should be reflected in the definition of the legislation and consequently the administrative procedures.

The mineral industry has unique characteristics when compared with other industries and requires adequately different treatment.[9] The extensive regulations may fall within the jurisdiction of different ministries (environment, mining, economy, energy, finance, trade, land use). If a deposit is situated in an area that because of existing land uses is incompatible with mining activities, the competent authority has the duty to decide which type of utilisation has the highest priority. The relevant proceedings should be governed by appropriate laws (e.g. mineral resource law, spatial planning law). Mineral resource law covers public law matters as well as important matters of private law, such as agreements with land owners or owners of neighbouring mines.

Furthermore, the appropriate regulations within mineral resource law can supply investors with comprehensive information concerning the role of the authorities, property rights, geological information, exploration and exploitation reports for authorities, costs and support for the investment in exploration and exploitation.

Main objective of a modern mineral resources law

The main aim of modern mining legislation is to set a legal framework which provides for orderly development of the mineral industry. For this purpose, such legislation must give investors the necessary incentives and guarantees, but it must also permit the government to give guidance to the industry and to provide mining policy that is in accord with broader economic and overall national policy. For these reasons, it must be considered within the general design of national policy and in the context of other relevant legislation, such as that which relates to the fields of environment, health and safety, taxation, customs, corporation law and investment incentives. Once the mineral rights (permissions, licenses or concessions) are bestowed and other approvals have

---

9   Cp. Tiess (2009), lc.

been granted, the responsible authority has to ensure the correct fulfilment of the regulations and the obligations agreed with the investors.

Mineral resources law must take into account features that are specific to the mining industry.[10] Transparent regulations are necessary to ensure not only legal certainty and effective procedures to minimize administrative costs, but also acceptable frame conditions with regard to taxes and fees. Against such a legislative background, prospecting, exploration, discovery and development of new deposits, as well as their exploitation (new investments) should be promoted. The law must provide a framework for safe and economic operation during the consecutive phases of the mining process. Modern mining requires clear and predictable decision-making criteria and emphasis on a standardised regime which meets business expectations. It requires measured administrative discretion by government agencies to minimise the chance of bureaucratic friction and prohibitive transaction costs. International competition between countries will only be enabled by efficient and transparent licensing authorities and procedures.

The system and acquisition of mineral rights

The system of the mineral rights in Europe originated in the 18th and 19th century. Laws such as the former Austrian Mining Act or the French Mining Code have produced a two-stage mineral legal system: **in general, mineral resources are owned either by the state or by the land owner.** Basically, a distinction can be made between minerals owned by the land owner, state-owned minerals and minerals "free for mining".

The assignment of property and ownership of realty is in principle effected by the judicial law provisions of a Civil Code. **Land owner minerals** are in this case directly subject to the civil law. That is, according to civil law, ownership of the raw materials is assigned to the respective land owners. For **authority** the entrepreneur needs power of disposal on the property for exploration and exploitation of (land owner) minerals and the utilisation of a property owned by others. The entrepreneur has to conclude a contract and usually has to pay a rental fee to the land owner or purchase the property.

However, the ownership of a property does not cover rights to **state-owned minerals**. In this case, the mineral law (or equivalent) overrules the Civil Code as a special law regulation so that the rights of property owners are limited to the extent that they cannot dispose of such minerals. The state attributes a national or economically significant interest to them. The explor-

---
10 Cp. Tiess (2009), lc.

ation and exploitation right must be purchased separately: for the exploration and exploitation of state-owned resources an exploration and production authorization (**mining concession**) is required in most cases. This is initially a government-assigned right for exploration and economic use of these raw materials. For any operational activities (exploration, extraction, processing) an additional regulatory approval is required for operating plans.

"Free for mining" minerals represent a further distinction[11]: The ownership of a property does not cover the "free for mining" minerals either, contrary to the civil law regulations. Since these raw materials are **not owned by anyone** the right to exploit them is not assigned to the realty. The Civil Code as a special law regulation is overruled by the mining law so that the rights of property owners are limited, with the result that the land owner cannot dispose of the "free for mining" minerals. (In Austria, the term "abandoned" minerals is commonly used). Ownership means in this case that the use of the minerals is entitled to no one. However, the government usually reserves the right to grant permission for the exploitation of the minerals by an administrative decision.

The key issues that the mining law should determine include: how mineral rights can be obtained, the nature of the rights that can be obtained, the obligations that are imposed, and the expiring of rights.

In most countries the state holds the disposal rights for state-owned minerals. Nevertheless, the state can grant disposal rights to entrepreneurs. It is important that this process is precisely defined in mining legislation, guaranteeing the legal certainty and predictability of the permitting process for the exploration and exploitation phase. Some issues are significant: how are exploration permits granted, are they limited or can they be expanded? If the exploration phase is successful, how great are the chances of subsequently obtaining an exploitation permit?

The exploration phase does not necessarily call for exclusive rights, but it is useful for the explorers to be registered and for their activity to be known to the government. For this reason, it is advisable to establish a non-exclusive exploration permit. In certain countries, this permit may be exclusive, in which case it is known as an "exclusive exploration license". In any event, the government should be able to create such zones in order to protect its own technical activities or those run by international bodies on its behalf.

The exploitation title must be of sufficient duration to allow for complete amortisation of the expenses of exploring, prospecting and equipping the mine

---

11 Cp. e. g. Article 3 (3) German Mining Law; Article 21, 28 Austrian Mining Law; Article 2 Norwegian Mining Law.

(including processing) and for repayment of the capital invested. In certain countries, there are two types of licenses – one that is broad in scope and of fairly long duration, and another one for simpler cases. The counterpart of exclusive rights is the obligation to work. Official supervision is needed to ascertain whether or not the work is actually being done and to make note of the results obtained. That obligation continues throughout life of a mine, but without undue interference with the management and financial equilibrium of the enterprise.

The **rights and obligations of a mining license holder** are seen as an important principle. Mining license holders are authorized, for example, to produce, operate and use for their own mining purposes mining facilities, company vehicles, surface mining equipment, operating equipment and the like.

However, the special entitlement of the mining license holders is matched by special obligations. Mining operators are obliged to make provisions for the protection of life, health and safety of people, the environment, the deposit and the land surface occupied during the mining activities, as well as the protection of surface use after the termination of mining. Before making use of the surface and the near-surface area of external (adjoining) properties for mining activities, the mining license holders have to obtain the agreement of the land owner. Realties and parts of them within the area covered by the exploitation license are lawfully held as mining areas.

An elementary question in view of the potential for foreign investors in the mining sector of a country is whether or not the particular state has provided an appropriate legal basis for mineral working. If that is the case, the law should make it clear if there is any difference between the rights afforded to domestic and to foreign investors, particularly with regard to questions of legal certainty. If possible, an attractive investment framework should be provided (e.g. on accounting, taxation law).

Technical and financial aspects

A very important feature of a modern mining law is the correct regulation of mining and environmentally relevant operations. The nature of mining activity is such that it can give rise to dangers and hazards which necessitate specific forms of regulation, particularly on environmental, health and safety aspects. Hence, as a rule there is an obligation to establish operating plans (exploration, exploitation and processing), which require an administrative permit in addition to the acquired mineral rights.

1.3 Requirements of a Mining Law

Most states dispose of the mineral rights (state-owned minerals) and recognise the obligation of **taxation**. In this way, an effective exploration process as well as efficient extraction of raw materials can be achieved. The extent of state intervention should be clearly defined by mineral resource law. In most cases, statutory regulations exist related to taxes and fees. Essential objectives are to guarantee economic and safe exploitation and a satisfactory mine closure. Moreover, exploitation of a deposit within a certain period of time (concession right) and optimal restoration must be achieved. Authorities are generally interested in swift exploitation of deposits, whereas mining companies are interested in a timely business start and a production rate that is in tune with market conditions. Generally it is possible to achieve a balance between both interests.

2 Albania

2.1 General Facts[1]

The national territory covers 28,748 km², and the inhabitants are 3,6 million. The population density is 126 inhabitants/km². The data on GDP per capita vary greatly (about 22 % of the average of the EU-27).

Constitutional Structure

The form of government is a parliamentary republic. The country is divided into three areas. 12 regions (quarks) are subdivided into 36 administrative districts which include 374 municipalities.

2.2 Production of Raw Materials

Table 3: Production Data – Albania (World Mining Data, Weber and Zsak)

Resources		2002	2003	2004	2005	2006	Change 02/06	Change 05/06
Iron	(t)	4.600	4.500	4.400	4.300	3.600	−21,74	−16,28
Chromium	(t)	9.000	9.500	9.800	12.000	15.000	66,67	25,00
Copper	(t)	0	0	600	1.700	400		−76,47
Gypsum	(t)	18.000	20.000	22.000	23.000	24.000	33,33	4,35
Salt	(t)	17.400	17.200	17.000	16.800	16.600	−4,60	−1,19
Sulphur	(t)	900	850	830	810	800	−11,11	−1,23
Brown Coal	(t)	20.300	19.000	15.000	20.000	12.000	−40,89	−40,00
Natural Gas	(Mm³)	9	9	10	11	11	22,22	0,00
Oil	(t)	330.000	340.000	350.000	370.000	518.686	57,18	40,19

---

[1] The general facts of the countries are based on the year 2006.

2.3 Normative Basics

2.3.1 Primary Legal Basics

The primary legal basics of mineral extraction activity is the Mining Law (ML)[2] No. 7796 of 1994, as amended by Law 6291 of 2004.

2.3.2 General Rules

Table 4: Structure of the Mining Law of Albania

Structure	
Chapter. I:	Name and Field of Use Articles 1–2
Chapter II:	General Principles Articles 3–7
Chapter III:	The State Authoritry and the Mining Rights Articles 8–13
Chapter IV:	Mineral Grouping Article 14
Chapter V:	Management Articles 15–17
Chapter VI:	State Mining Enterprises Articles 18–20
Chapter VII:	Mining Rights for Mineral Groups 1,2 & 3 Prospection Permit Articles 21–29 The Exploration Permit Articles 30–43 The Exploitation Permit Articles 44–62
Chapter VIII:	Mining Rights for Mineral Group 4 and Construction Materials Prospection and Exploration Permit Articles 63–64 Exploitation Permit Articles 66–75
Chapter IX:	Mining Rights for Mineral Groups 5 & 6 Articles 76–87
Chapter X:	Relations between the Mining Permit Possessor and the Owner of the Land Article 88
Chapter XI:	General Obligations of Possessors of the Mining Permit Articles 89–99
Chapter XII:	Specific Benefits to Promote Private Investments Articles 100–101

---

2 Regarding the term "Mining Law": It is observed that in different countries there are often different terms used, e. g. Mines and Quarries Law (Cyprus), Mining Act (Czech Republic), Mining Code (France [Code Minier] Greece), Mineral Development Act (Ireland), Law on Subsoil (Latvia, Russian Federation) or Subsoil Code (Moldavia), Underground Law (Latvia). To get an overall "comparable picture" the term 'Mining Law' (abbreviated as ML) will be used in all countries as "synonym term".

Structure	
Chapter. XIII:	The Register of Mining Permits Articles 102–104
Chapter XIV	Disciplinary Measures adopted in Cases of Exercise of Mining Activities without Valid Mining Permits, the Right of the Minister to cancel the Mining Permit Articles 105–108
Chapter XV:	Miscellanous Articles 109–113

The Mining Law covers the following minerals categories:

a) First Group: Metal minerals
b) Second Group: Non-metal minerals
c) Third Group: Coals and bitumen
d) Fourth Group: Minerals and construction
e) Fifth Group: Precious stones
f) Sixth Group: Half-precious stones

Ownership of mineral rights

The field of this law is the entire territory of the Republic of Albania including the submarine and subterranean parts that are under its jurisdiction (Article 2 ML).

Acquiring mineral rights

The state has responsibility about issuing mineral rights to natural persons and legal entities.[3] According to Article 15 ML the responsible minister is the competent authority for issuing mineral rights.

2.3.3 Issuing of Permits

Mineral category 1, 2, (Chap. VII, mining rights for groups 1, 2 & 3)

Prospecting permits Articles 21–29 ML

The prospecting permit ("search permit") is regulated by Articles 21–29. ML. Based on an application of a natural person or legal entity the competent authority issues a prospecting permission.

The permitted prospecting area should not exceed 400 m², must be unfragmented and should not cover any area, which is object of another prospecting permit granted in compliance with this law (Article 22 ML). The maximal time

---

[3] The term "mineral right" is discussed in section 1.3.

limit for a prospecting permit is one year and is not object of an extension. The application for a prospecting permit must include (Article 25 ML):

a) Name and address of the applicant;
b) Specifications of financial resources and technical capacities needed to the permit applicant as well as his experience in the field of mining industry;
c) A description of the prospecting area including a map;
d) Specifications of the proposed search program, the proposed methodology, an estimation of costs and the time limits of program execution.

Within 60 days from the application receipt, the Minister must notify the applicant about his decision with regard to the application for prospecting permit (Article 26). The prospecting permit grants the following rights and obligations (Article 27 and 28 ML). The possessor of a prospecting permit is entitled to enter in the permitted area with personnel, means, machineries and equipment necessary or indispensable for the conduct of the search of minerals on or under the ground. The possessor of a prospecting permit is obliged to report to the competent authority in writing about the state of all minerals having an economic interest (Article 28).

Exploration permit (Articles 30–43 ML)

Articles 30–43 regulate the issuing of the exploration permit. Considering the technical and financial means/resources the competent authority issues the exploration permit. For this purpose the applicant must prepare data and documents according to Article 31 ML, this documentation includes:

a) Name and address of the applicant;
b) The applicants experience in the exploration of minerals;
c) The existing financial and technical resources needed by the applicant to obtain the permit;
d) The proposed area for the permit including a map of the area;
e) The proposed methodology of exploration;

The proposed work program and a work-plan must include deadlines and the estimated expenses allocated for its implementation. An exploration permit guarantees the possessor the exclusive right to discover the specified minerals in the permitted area (Article 30 ML). The competent authority within 60 days from application receipt must notify the applicant about his decision (Article 33 ML).

The initial time limit of the exploration permit is two years and may be extended by one year with a maximum of three extensions if requested by the permit possessor (Article 34 ML). The application for the extension of the exploration permit must be submitted at least 30 days prior to the expiration of the current permit. These time limit extensions have to be granted if the competent authority deems that the permit possessor has fulfilled the financial obligations in conformity with the permit issued and submit a convincing work program for further continuation of the discovery.

Within 90 days after the effective date of the permit, its possessor must commence works in the field (Article 35 ML). The maximal permitted area of an exploration permit is 200 km² and is unfragmented (Article 36 ML). A possessor of the exploration permit may hold more than one permit simultaneously (Article 37 ML).

According to Article 38 ML the possessor of the exploration permit is obliged to/must leave parts of the permitted exploration area progressively in conformity with the time limits. By the end of the initial time limit, the possessor must leave at least 40 % of the initial permitted area. The possessor of the exploration permit must keep a complete and detailed record of the work conducted for the discovery of minerals and other operations in conformity with the exploration permit. These records are at any moment valid for inspection in Albania by the competent authorities (Article 42 ML). The possessor of the exploration permit is entitled before the expiration of his permit to apply for a mining permit (Article 43).

Exploitation permit (Articles 44–62 ML)

Articles 44–62 ML regulate the provisions in terms of exploitation permit (see table 4). The competent authority grants natural persons or legal entities by considering financial and technical resources and mining experience an exploitation permit. This permit guarantees the exclusive right to use one or more minerals specified in the permitted area (Article 44).

Article 45 ML regulates the content of the exploitation application. The application for an exploitation permit must include:

a) Name and address of the applicant;
b) Specification of the area for which the exploitation permit is requested including a map;
c) Specification of the mineral or minerals in groups 1, 2 or 3 for which the exploitation permit is requested;

When the applicant of the exploitation permit is not possessor of the exploration permit, the application must include complete details of the financial and technical capacity of the applicant and information regarding his experience in the mining field (Article 46 ML).

The maximal permitted area of a mining exploitation permit is 15 km² and is unfragmented. A possessor of the mining exploitation permit can hold more than one permit simultaneously (Article 48 ML).

The time limit of a mining exploitation permit is up to twenty years, and is subject of four extensions up to five years each, if the permit possessor requests these extensions in a time no later than one year prior to the expiration data of the previous time limit (Article 49 ML).

Rights and duties

The exploitation permit allows the possessor rights and duties. These are settled in Articles 52–59 ML. According to Article 52 the mining exploitation permit grants the following rights to the possessor:

a) To execute necessary actions to implement effectively the mining operations in his permitted area of exploitation;
b) To mine and remove from the permitted area the mineral or minerals for which the exploitation permit has been issued;
c) To install and operate within the permitted area benefication units, casting plants, refineries and other factories that increase the value of the mined minerals in his permitted area of exploitation.

The possessor of the exploitation permit has the right without authorization to sell in Albania or to use for export the produced minerals and/or final products (Article 55 ML). The possessor of the exploitation permit must submit the financial and technical report of all operations executed in compliance with the mining law and must make available these reports for inspection at any time by the competent authority (Article 56 ML).

The possessor of the exploitation permit must submit to the competent authority an annual working program, the budget and the plan of production of his permitted area of exploitation (Article 59 ML).

When the competent authority considers that the possessor of the exploitation permit is using harmful practices regarding the minerals and/or their treatment or is exploiting the mines using non optimal methods, it must notify the permit possessor and request within a certain time limit the necessary explanations (Article 60 ML).

According to Article 62 ML the exploitation permit possessor has to notify the competent authority:

a) One year in advance it is proposed to stop the production of a mine in the above mentioned area; b) 180 days in advance if proposed to suspend the production of such mine; c) 90 days in advance if proposed to reduce the production of such mine with more than 10% of the planned production in the approved production plan.

Prospection and exploration permit (Article 63 and 64 ML)

Article 63 regulates the prospection and exploration permit. Article 63 ML requires different information.

An exploration permit has a maximum time limit two years and a maximum surface of 30 km² (Article 63 ML). The extension of the permit depends on the specifications of the activity the person wants to be executed. In the case of an exploration permit, the person must pay to the state the equivalent of 100 SFr/km² of the permitted area.

Exploitation permit (Article 66–75 ML)

Article 70 ML requires the following:
a) Name and address of the applicant;
b) To define the minerals for which the permit is applied.
c) To specify the initial time limit of the permit;
d) To provide details of financial resources applicable to applicants of the permit with the purpose of expenses payment of exploitation of minerals and the construction materials;

The competent authority must notify on its decision on the application for a quarry permit within 30 days from its receipt (Article 71 ML). An extraction permit according to Article 69b ML is an exclusive permit to exploit construction materials on payment (that the competent authority will determine in the issued permit; the amount of this payment is determined in the regulation). The initial maximum time limit of the quarry permit is ten years, subject to renewal every year (one after the other) according to application of the possessor, which has to be submitted at least thirty days prior of expiration of the previous period.

Closure phase for all mineral categories

In cases when the exploitation permit has expired and the area pertaining to the permit has been abandoned, the possessor of the exploitation permit prior to expiration of the permit or abandonment of the mining area must rehabilitate and restore it to its previous natural state within a reasonable period. This action is considered completed upon approval by the competent authority (Article 99 ML).

2.3.3.1 Authorities

According to Article 15 ML the main responsible authority for mining is the Minister of Economy, Trade and Energy (General Directorate of Mines). It will be represented from the General Directorate of Mines (Article 16 ML).

The competent authority has the right to cancel any mining permit to the possessor in the following cases: When committing violations of the Mining Law; interrupts the work in a mine for more than 180 consecutive days (Article 106 ML).

2.3.3.2 Fees and Taxes

The exploration permit possessor of the mineral categories 1–3 must pay to the state an annual payment for his exploration area. The total annual payment is equivalent to 300 SFr (e.g. € 250; Article 40 ML).

Regarding mineral category 4, the quarry exploration permit possessor must pay to the state a total payment of a 100 SFr/km² of the permitted area (Article 63 ML).

According to Article 50 ML the exploitation permit possessor must pay to the state an annual payment for the permitted area of exploitation. The annual payment for a mining exploitation permit has a minimum value of the equivalent in all of 3.000 SFr and the maximum value of the equivalent of 10.000 SFr/km² (e.g. € 2.500 – € 8.400).

The possessor of the exploitation permit (mineral categories 1–3) no later than 15 days after the end of each calendar month must pay to the state in equivalent of Swiss Franks a rent (royalty) for the mining property equal to 2% of the market value of the overall sold mineral quantity to consumers during this calendar month and produced by the permitted area (Article 51 ML).

The quarry permit possessor must pay to the state a total annual payment equivalent of 1.500 SFr/km² of the permitted area of exploitation (Article 73 ML).

The quarry permit possessor, no later than 15 days after the end of each calendar month, must pay to the state a mining property rent (royalty), in equivalent SFr, 2% of the market value for the gross quantity of mineral sold to consumers during that calendar month and produced by this permitted area (Article 74).

2.3.4 Additional Legal Basics

Table 5: List of Acts significant for raw materials – Albania.

Name of Law	No. of Law and Year of Issuing
Regulation on the Plans of Rehabilitation of the Areas damaged by Mining Companies	Law No. 3 of 2006
Regulation on the Establishment of the National Agency of Natural Resources	Law No. 547 of 2006
Law on Environmental Protection	Law No. 8934 of 2002
Law on Environmental Impact Assessment	Law No. 8990 of 2003
Law on Protection of Air from Pollution	Law No. 8897 of 2002
Law concerning the Environmental Treatment of Polluted Waters	Law No. 9115 of 2003
Law on Water Resources	Law No. 8093 of 1996
Civil Code	Law No. 7850 of 1994
Code of Administrative Procedures	Law No. 8485 of 1999
Law on Income Tax	Law No. 8438 of 1998
Law on Competition Protection	Law No. 9121 of 2003
Law on Foreign Investments	Law No. 7764 of 1993

Environmental protection Law

Environmental protection constitutes an essential factor for the development of the society and the nation in general, and has these main strategic elements: prevention and reduction of pollution of water, air, ground, of any kind; rational exploitation of natural resources, the avoidance of over exploitation; the ecological restoration of areas damaged by human activities or natural destructive phenomena; preservation of ecological equilibrium; and life quality maintenance and improvement (Article 1). Article 18c covers the (environmental) licence concerning exploration, extraction, and exploitation of 'subsoil' resources.

Law on Water Resources

The purpose of the Law on Water Resources is to ensure conservation, development and utilization of water reserves in a way as rational as possible, vital for life and social-economical development of the country; to ensure the right distribution of water reserves according to utilization and their effective management and to ensure protection of water reservers from pollution, abuse and overconsumption beyond the actual needs (Article 1). Permits related to minerals extraction from the beds and banks of rivers, streams, lakes is issued by the water authorities (Article 20).

3 Austria

3.1 General Facts

The national territory covers 84,000 km² and the inhabitants are 8.3 million. The population density is about 100/km². The GDP/per capita amounts to $ 39.400 (2009 estimation).

Constitutional Structure

The form of government is a democratic republic. Austria is divided into a national level, provincial level (nine federal states), district level (84 districts) and municipal levels.

3.2 Production of Raw Materials

Table 6: Production Data – Austria (World Mining Data, Weber and Zsak, 2008)

Resources		2002	2003	2004	2005	2006	Change 02/06	Change 05/06
Iron	(t)	621.363	679.932	604.614	665.344	669.438	7,74	0,62
Tungsten	(t)	2.242	2.250	2.240	2.365	2.001	−10,75	−15,39
Gypsum	(t)	969.202	1.003.550	1.038.127	1.017.194	1.071.452	10,55	5,33
Kaolin	(t)	50.908	100.331	104.986	55.508	51.900	1,95	−6,50
Magnesite	(t)	728.235	766.525	715.459	693.754	769.188	5,62	10,87
Salt	(t)	964.882	1.028.273	1.030.234	1.024.090	807.278	−16,33	−21,17
Sulphur	(t)	9.444	10.400	10.705	8.458	10.166	7,65	20,19
Talc	(t)	138.195	137.596	136.305	166.569	159.447	15,38	−4,28
Brown Coal	(t)	1.411.824	1.152.389	235.397	6.168	7.854	−99,44	27,33
Natural Gas	(Mm³)	2.015	2.030	2.011	1.654	1.765	−12,41	6,71
Oil	(t)	1.032.204	1.014.716	981.588	854.775	856.270	−17,04	0,17
Oil Shales	(t)	336	432	248	0	287	−14,58	

Table 7: Aggregates Production – Annual Statistics/Austria, 22 April 2008 (UEPG 2008)

Sand & Gravel (1)		Crushed rock (2)		Marine Aggregates		Recycled Aggregates (3)		Manufactured Aggregates (4)	
2005	2006	2005	2006	2005	2006	2005	2006	2005	2006
66,0	66,0	32,0	32,0	n.a.	0,0	3,5	3,5	3,0	n.a.

(1) Sand and Gravel: sold production including crushed gravel
(2) Crushed rock: sold production (excluding crushed gravel)
(3) Recycled Aggregates: materials coming from construction and demolition waste used in aggregates market
(4) Manufactured aggregates include blast-furnace-slag, electric-arc-furnace-slag, incinerator bottom ash (IBA), pulverised fuel ash (PFA)

3.3 Normative Basics

3.3.1 Primary Legal Basics

The primary legal basics of mineral extraction activity is the Mining Law ('Mineralrohstoffgesetz') No. 38 of 1999 as amended by Law 115 of 2009.

3.3.1.1 General Rules

Table 8: Structure of the Mining Law ('Mineralrohstoffgesetz') – Austria

I. Main Part. General regulations
Articles 1 to 5
II. Main Part. Exploration for raw materials
Articles 6 and 7
III. Main Part. Exploration and exploitation of free for mining minerals
I. Section.
Exploration permit Articles 8 and 9
Grant of exploration permits Articles 10 and 11
Inadmissibility of amendments for applications Article 12
Extension of validity period of exploration permits Article 13
Transfer of exploration permits Article 14
Expiration of exploration permits Articles 15 and 16
Work plan Articles 17 and 18
Change of work plan Article 19
Exploration report Article 20

Transfer of ownership during exploration of accruing free for mining minerals Article 21

II. Section.

Mining permits Articles 22 and 23
"Grubenmaße" Articles 24 to 32
"Überscharen" Articles 33 to
Registration in the catalogue of mining property Articles 40 to 43
Operating obligation in "Grubenmaßen und Überscharen" Articles 44 to 50
Transfer of exploration permits and relinquishment of the operation Articles 51 to 53
Deactivation of mining permits Articles 54 to 65
Withdrawal of mining permits Articles 66 and 67

IV. Main Part. Exploration and exploitation of state owned mineral raw materials

I. Section.

General Article 68
Relinquishment of rights Articles 69 and 70)

II. Section

Work plan Articles 71 and 72

III. Section.

Exploitation field Articles 73 to 79

V. Main Part. Extraction of surface land owner minerals

Extraction plan – content Article 80
Status of involved parties Article 81
Extraction plan – land use planning Article 82
Extraction plan for land owner minerals – additional permit preconditions Article 83
Mining permit holder Article 84
Closure of exploitation Article 85

VI. Main Part. Storage of hydrocarbon in non-hydrocarbon geological formations

I. Section. Search and exploration of non hydrocarbon geological formations Articles 86 to 88

II Section. Permit for storage Articles 89 to 96

VII. Main Part. Assertion of mining permit holder

I. Section. General regulations Articles 97 to 101

II. Section. Special rights of the mining permit holder Articles 102 to 107

III. Section. Special duties of the mining permit holder Articles 108 to 111

IV. Section. Extraction plans, mining facilities, mining equipment Articles 112 to 124

V. Section. Responsible persons Articles 125 to 142

VI. Section. Mining permit holder Article 143

VII. Mining permit holder Article 144

VIII. Section. Liability for cash benefit Article 145

IX. Section. Exclusion of a separate transaction Article 146

VIII. Main Part. Mining and property

I. Section.
Relinquishment of property Articles 147 to 151

II. Section.
Relinquishment of the use of surface water Article 152

III. Section.
Mining areas Articles 153 and 154
Notification to the cadastral register court Article 155
Rejection of a construction permit Articles 156 to 158

IV. Section.
Securing the surface use after finalisation of the mining operation Article 159

V. Section.
Mining damages Articles 160 to 169

IX. Main Part. Authorities

I. Section.
Jurisdiction of the authorities Articles 170 to 172

II. Section.
Functions of the authorities Articles 173 to 174

III. Section.
Cooperation of the authorities with other bodies Article 176

IV. Section.
Supervision power, information and toleration duties Article 177

V. Section.
General directive power of the authorities Articles 178 to 180

VI. Section
Enactment of regulations for protective measurements for mining Article 181
Enactment of regulations for protective measurements for processing Article 182

VII. Section.
Application of the protective law for employees Article 183

VII. Section.
Relation to other regulations for realisation of operations by third-party contractors Article 184

IX. Section.
Marking and overall plans Article 185

X. Main Part. Costs

Article 186

XI. Main Part. Central office for rescue and gas defence

Articles 187 to 189

XII Main Part. Mining advisory board	
Article 190	
XIII. Main Part. Charges for "Freischurf- und Maßengebühren"	
Article 191	
XIV. Main Part. Award	
Article 192	
XV. Main Part. Sanctions	
Article 193	
XVI. Main Part. Annulment, Transitional and Final Provision	
Articles 194 to 224	

The Mining Law distinguishes between three mineral categories:
- "Free for mining" raw materials:
  - "Free for mining" raw materials neither belong to the State nor to the ground owner. Out of these raw materials metals, gypsum and anhydrite, graphite, talc, coal, magnesite and limestone ($CaCO_3$ quotas of more than 95 %) are extracted.
- State-owned raw materials:
  - State-owned raw materials are owned by the State. Rock salt, hydrocarbons as well as raw material which contain uranium or thorium are State-owned raw materials.
- Land owner raw materials:
  - Land owner raw materials are owned by the landowner. These are all raw materials which are not "free for mining" and State-owned raw materials (such as dolomite, quartzite, bentonite, diatomite, asbestos, mica, feldspar, marl, granite).

Ownership of mineral rights

The ownership rights of raw materials are determined in the Mining Law. The state has the responsibility regarding issuing mineral rights. A mining right is the right to explore and/or exploit mineral resources for the purposes of industry and trade. The mineral right holder is a legal or natural person who has obtained the mining right by concession in accordance with the qualified provisions of the law. According to Article 1 (14) Mining Law, a mining right can be:
- **Exploration right**:
  - Exploration permit
  - Right of the State to explore State owned raw materials
- **Exploitation right**:
  - Free for mining raw materials: mining permit

- State-owned raw materials: right of the State to extract State owned raw materials. Approved exploitation plan for the extraction of State owned raw material with the exception of hydrocarbons
- Land owner raw materials: Exploitation plan (equivalent of a mining right

The Mining Law distinguishes between the exploration phase and the exploitation phase concerning the acquirement of mineral rights. The exploration phase is divided into: search for minerals (prospection); examination of the occurrence of raw material in order to determine if it is worth extracting (exploration).

Table 9: Mineral Rights according to the Mining Law

Category of Mineral Rights	Description of Mineral Rights	Minerals Covered by Rights
"Free for mining" raw materials	A raw material excluded from the right of disposal of the landowner, which may be explored for and extracted by any person fulfilling certain legal requirements.	1. All mineral raw materials from which iron, manganese, chromium, molybdenum, tungsten, vanadium, titanium, zirconium, cobalt, nickel, copper, silver, gold, platinum and platinous metals, zinc, mercury, lead, tin, bismuth, antimony, arsenic, sulphur, aluminium, beryllium, lithium, rare earths associated with it. 2. Gypsum, anhydrite, barite, fluorite, graphite, china clay, and leukophyllite. 3. Magnesite. 4. All types of coal and oil shale.
State-owned raw materials	Mineral raw materials which are the property of the State.	1. Rock salt and all other salts associated with it. 2. Hydrocarbons. 3. Uranium and thorium bearing raw materials.
Landowner raw materials	Minerals belonging to the land owner.	Gravel and sand, etc.

"Free for mining" raw materials require a mining permit. This allows the holder of the license to extract and to acquire "free for mining" raw (Article 22 ML). Mining rights are granted for mining claims ("Grubenmaß"). A mining claim is not limited to the depth and its cut surface in the projection level of the national

surveying system builds a flat rectangle with an area of 48 000 m². Mining rights for mining fields are conferred by the competent authority, if a (judicial) person applies for them. Conditions for the permit are:

- The occurrence of "free for mining" raw material has to be economically viable
- Other permits must not be opposed to the mining right
- Consideration of the public interest (nature conservation, land use planning, tourism, environmental protection, water management, (rail) traffic, national defence)

State-owned raw materials require the approval of an exploitation field. An exploitation field is a space which is not limited to the depth and its cut surface in the projection level of the national surveying system is a flat polygon. If State-owned raw materials other than hydrocarbons (i. e. rock salt) occur, the area must not be larger than 1 km². If the required conditions are fulfilled, the exploitation field has to be approved by the authority (Article 74 ML).

3.3.1.2 Issuing of Permits

Prospecting as well as exploration is regulated in the Mining Law with different provisions for each raw material category:

Prospecting (Articles 6 and 7 Mining Law)

- "Free for mining" and landowner raw materials:
  Due to the fact that in the first phase of exploration it is not clear which raw material will be found, it is not possible to differentiate between "free for mining" and landowner raw material. For the first phase of exploration it is only necessary to make an announcement to the authority and to present a report with the results at the end of each year.

- State-owned raw material:
  The exploration of State-owned raw materials is regulated separately. For legal reasons exploration is reserved to the State.

Exploration (examination of an occurrence in order to determine if it is worth extracting)

- "Free for mining" raw materials
  In order to examine natural occurrences of "free for mining" raw materials, an exploration permit ("Schurfberechtigung") is required (ML).

- State-owned raw materials
  The exploration of State owned raw materials is for legal reasons reserved to the State (Article 68 ML); the authority has to approve exploration activities. The exercise of the rights concerning rock salt is reserved to the Austrian Salinen AG.

Exploitation

As already mentioned, mining permits represent legal entitlements. In order to exercise these rights other authorisations and permits are also required. In order to authorise the mineral exploitation concerning "free for mining" as well as State-owned raw materials (with the exception of hydrocarbons) an exploitation plan is required. For surface and underground exploitation of *landowner raw materials* an exploitation plan is required; this plan includes additional regulations compared to the exploitation plan for "free for mining" and State-owned raw materials.

Application requirements – Exploitation plan

The operator has to declare his plans to (explore and) extract raw materials to the authority; the announcement has to contain an exploitation plan with the following content according to Article 113 ML[1]:

- Planning period of time of the mining intention
- A description of the planned extraction and the transport of the raw material
- The planned safety measures
- Information about the emissions expected due to the extraction and statements to reduction
- Description of the measures taken to protect the surface and to protect the surface utilisation after terminating the extraction as well as information about the costs for these measures
- Information about the expected utilisation of the open cast area
- An exploitation plan which should include the following (only concerning "free for mining" raw materials and State-owned raw materials):
  o Layout plans of the mining areas as well as the measures which have to be taken in order to guarantee the safety of the surface and the safety of the surface after terminating the mining activity

The permits of exploitation plans by the mining authority requires the following conditions (Article 116 ML)[2]:

---

1 According to Article 113 ML; Article 113 (1) applies to all raw material categories.
2 Article 116 ML applies to all raw material categories.

- The mining activities which are listed in the exploitation plan must be approved by a mining permit.
- No destructive exploitation of a deposit (i.e. the extraction has to correspond with the technical, economical and security technical requirements).
- Dealing with the surface has to be economical and careful.
- Measures to protect the surface utilisation after terminating the mining activities need to be planned.
- Avoidance of emissions according to best technical standards is carried out.
- No harm of life or health of people (as well as no unreasonable pestering)
- No unreasonable harm to environment.
- The production of waste should be avoided according to best technical standards.

Consultation process

The Mining Law states that, when considering an application for an extraction permit (for "free for mining" mineral raw materials or for landowner's mineral raw materials), the owner of the land must be involved in the process. The provincial authority also becomes a (legal/formal) consultee in so far as the application relates to land use planning, protection of the nature/environment, tourism or other aspects. One aim of this process is to ensure that the "public interest" is taken into account. This can also mean consulting other relevant authorities (e.g. in relation to transport, the environment, water). Exploitation plans (see above) may only be approved if the proposed measures are sufficient to protect surrounding property, the neighbours, the safety of people, and the environment as addressed in conditions.

Duration of Permits

Exploration permits are granted for fixed periods of time; their period of validity may be extended under certain circumstances. Extraction permits for mineral raw materials for which no ownership rights exist ("free for mining minerals") are open-ended. However, the issue of a permit requires that extraction operations actually take place. If this requirement is not met, the extraction permit may be withdrawn.

In the case of land owner's minerals, the period of validity of the extraction permit is decided in accordance with civil legal considerations. In most cases permits are valid only for a few years whilst processing and operating permits for mining installations usually remain valid for the period of the operations.

The exploration permit is conferred according to Article 13 ML for the period of the current legal year and the following four legal years. The period of validity can be prolonged about five years, if prospecting and exploration are pursued.

There is operating liability for mine fields (Article 44 ML: The extraction of "free for mining" raw materials has to begin within two years after the exploitation permit has been legally conferred. According to Article 45 Mining Law, the entitled person is obliged to extract at least four months per year.

Exploitation plan for the extraction of "free for mining" and State-owned raw material: An exploitation plan for five years must be set up (Article 112 ML). Under certain circumstances a shortening or extension of the time period is possible. If there is an interruption of more than five years, an exploitation plan is also required.

Closure phase

Under the Mining Law, operators are required to notify the competent authority of their intention to cease mining and to decommission the plant. The mine closure plan must include (Article 112 [2] and Article 114 ML):

- An exact demonstration of the technical realisation of the closing and securing operations.
- Documents dealing with the securing of the surface in the interest concerning security of persons and things.
- Documents dealing with the required precaution in order to make the surface usable again.
- Documents describing the significant geologic deposits.
- A register of the mining maps.

The mine closure plan has to be approved by the competent authority (Article 117 ML). Securing of the surface use after terminating the mining operations (Article 159 ML). The entitled person has to take measures to secure the surface use after terminating the mining operations. External properties which have been used for mining operations must be put in their former or similar condition.

Restoration

Guaranteeing Funding for Restoration: the responsibility for carrying out and paying for restoration rests with the operator: The competent authority must ensure that restoration takes place as agreed. Normal practice is for the operator to deposit an agreed amount of money with a bank sufficient to secure restoration of the site. The relevant authority will check with the bank to ensure that this has been undertaken.

3.3.1.3 Authorities

The main responsible authority for mining is the Minister of Economy, Family and Youth (General Directorate Energy and Mining).

The conditions for approval of exploitation plans and for the authorisation of mining installations are laid down in the Mining Act. The competent authority will first wish to ensure that the activities covered by the exploitation plan are covered by a mining permit, that the applicant has the financial background to carry out the activities of restoration and aftercare, that the proposed conditions are sufficient to ensure health and safety of people, and to protect the environment, and neighbouring property and other mineral deposits and land surfaces.

In order to exercise the right to extract, the operator must have the consent of the landowner. He does not necessarily have to own the land. If the operator does not own the land, then it is normal for an indemnity to be paid to the landowner as compensation for use and damage of the land. If the two parties are unable to agree on a figure, they can request the Mining Authority to decide on an appropriate amount.

3.3.1.4 Fees and Taxes

Fees must be paid for office actions (conveyance of mining areas (Article 158 ML), general office actions (Article 186 ML), measures which are taken by the authorities (Articles 178 and 179 [5] ML) and the mine rescue and gas protection organisation (Article 187 ML). The permit holder of a right for exploration and extraction of "free for mining" minerals must pay a mining fee in which the area of entitlement is located. Fees are to be paid for the exploration permit of the mining area (see Table 10) for "free for mining" minerals (Article 191 ML); and surfaces interest, field interest, storage interest and conveyor interest for the State owned minerals (Article 69 [1] ML).

Table 10: Fees according to the mineral rights (for each calendar year)

Mining right	Kind of the fee	Amount of the fee
Exploration permit for "free for mining" raw materials	Exploration Permit	8,72 Euro
Mineral right for "free for mining" raw materials	Fee for the mine dimension	26 Euro
	Fee for the "Überschar"	13 Euro

Compensation must be paid for the use of land to drain mine waters (Article 106 [4] Mining Act), use of property for mining operations (Article 148 Mining Act), refusal of a building permit interfering with mining activities (Article 156 Mining Act), securing of the surface use after terminating the mining operations (Article 159 Mining Act), and mining damages (Articles 160, 161, 164, 168 and 169 Mining Act).

Penalties of unauthorised or damaging activities which are connected with mining are provided in the Mining Law (Article 193 Mining Act).

3.3.2 Additional Legal Basics

Table 11: List of Acts significant for Raw Materials – Austria

Name of Law	No. of Law and Year of Issuing
Environmental Impact Assessment Act	Law No. 89 of 2000 as amended by Law No. 2 of 2008
Emmission Control Act – Air	No. 115 of 1997 as amended by Law No. 70 of 2007)
Waste Management Law	Law No. 200 of 102, as amended by Law No. 54/2002
Trade regulation	Law No. 194 of 1994, as amended by Law No. 68 of 2008
Forestry Act	Law No. 450 of 1975, as amended by Law No. 55 of 2007
Water Act	Law No. 215 of 1959, as amended by Law No. 123 of 2006
General Administrative Procedure Act	1991
Civil Code (Allgemeines Bürgerliches Gesetzbuch)	
Environmental Information Law	495 of 1993
Nature Conservation Law and Land Use Planning Law enacted by the provinces	

Environmental Impact Assessment Act

The combined authorisation for developments falling under the Environmental Impact Assessment Act is conducted by the provincial government for the area. Environmental effects are identified, described and evaluated by the environmental impact assessment in the planning stage. Of great importance is the integration of the general public environmental impact assessment process.

Water Act

The Federal Water Act defines many prohibitions, rules and approval requirements which apply in general to all branches of industry, including mining. The requirement to obtain a separate permit under the Water Law is waived for certain projects subject to the Mining Act and regulations in order to streamline the administrative process. In such cases, water issues are dealt with the procedures required under mining legislation. Standard emission limits, and specific limits for certain industrial sectors, have been set for discharges into water bodies. If the ground-water level is influenced by any mining activities, a permit according to the water right is required.

Forestry Act

The clean air provisions of the Forestry Act are particularly important for the exploration, extraction, preparation, refining and processing of raw materials. The Forestry Authorities are responsible for implementing the Forestry Law. Certain listed installations which emit air pollution likely to cause damage to forests require approval. In order to simplify administration, such installations requiring a permit under the Mining Act, are not required to obtain a separate forestry law permit, unless the forests in question are protected or act as a protective barrier against avalanches. If the requirement is waived, the relevant provisions of the Mining Law come into force.

Emission Control Act – Air

The compliance with formalities according to the Emission Control Act – Air ("Immissionsschutzgesetz-Luft (IG-L)") for clean air permits is the responsibility of the mining authority.

Waste Management Act

The Waste Management Act is implemented in the first instance by the head of the provincial government. Decisions can be conferred by the head of the provincial government on the district administrative authorities.

Nature Conservation Act ("Naturschutzgesetz")

The Nature Conservation Act of the provinces is implemented by the Nature Conservation Authorities (the District Administrative Authorities and the Provincial Government).

4 Belarus

4.1 General Facts

The national territory covers 207.600 km² and the inhabitants are 9.689.800. The population density is 46,7/km². GDP per capita for 2009 is estimated at $ 5.165.

Constitutional Structure

The form of government is a presidential republic. Its Capital is Minsk. Belarus is divided into 6 provinces (voblasts) which are named after their administrative centres – cities; each of them has a provincial legislative authority (oblsovet) and a provincial executive authority. Voblasts are subdivided into districts (raions). As of 2002 there are 6 oblasts, and 102 towns.

4.2 Production of Raw Materials

Table 12: Production Data – Belarus (World Mining Data, Weber and Zsak, 2008)

Resources		2002	2003	2004	2005	2006	Change 02/06	Change 05/06
Potash	(t)	3.790.000	4.230.000	4.600.000	4.844.000	4.605.000	21,50	−4,93
Salt	(t)	136.900	154.300	188.297	183.900	180.000	31,48	−2,12
Nat. Gas	(Mm³)	246	254	245	228	225	−8,54	−1,32
Oil	(t)	1.846.000	1.820.000	1.804.100	1.785.020	1.800.000	−2,49	0,84

Table 13: Production of lime – Belarus. (Metric tons), (USGS)

Belarus	2003	2004	2005	2006
Lime	658	727	785	853

4.3 Normative Basics

4.3.1 Primary Legal Basics

The main legal basics is the Mining Law ("Subsoil Code"), Law No. 103-Z of 1999.

4.3.1.1 General Rules

Table 14: Structure of the Mining Law – Belarus

Section I	General provisions
Section II	Regulation of use and protection of subsoil
Section III	Use of subsoil
Section IV	Rational use and protection of subsoil
Section V	State registration
Section VI	Charges for the use of subsoil
Section VII	Security of exploitation of subsoil
Section VIII	Control and supervision
Section IX	Liability for infringements, compensation, dispute settlements
Section X	International agreements
Section XI	Conclusive provisions

Ownership of mineral rights

The ownership of mineral rights is the exclusive property of the state (Article 3 ML).

4.3.1.2 Issuing of Permits

The basic principles of subsoil use (Article 10 ML) are:
- Rational and integrated use of mineral resources and their protection;
- Paying for subsoil use; licensing procedure for obtaining the right to subsoil use primarily on a competitive basis; combination of national and international interests in the use and protection of natural resources;
- Stimulation subsoil users for activities to improve the use of mineral resources.

Use of mineral resources is temporary. Duration limit of extracting minerals is up to 20 years. If necessary, the period can be extended for a period, at the request of the user of mineral resources (Article 13 ML).

The state system of licensing of subsoil use is a single procedure for granting permits including economical, environmental requirements. Issuing of subsoil use permit rests with the Minister of Natural Resources and Environment. The permit for using mineral resources is a document certifying the right of its owner to use subsoil areas within certain limits (Article 21 ML).

Granting permits (Article 22 ML)

Granting permits for subsoil use is carried out on the basis of decisions of the Minister of Natural Resources and Environment. Grant of permit is done primarily on a competitive basis. Information on upcoming auctions and their results will be published. The main criteria during the tender or auction are:

- The scientific and technical level programs for the use of mineral resources;
- The completeness of extraction of minerals and waste management;
- The contribution to the socio-economic development of the territory; timing of related programs; environmental performance.

Preferential right to a tender or auction has an operator who proposed an exploration area.

The right of subsoil use cannot be transferred to third parties, except: 1) changes in the legal form of enterprise; 2) reorganization of the enterprise (Article 23 ML).

The right of subsoil use is terminated if the need arises for public purposes; reorganization of a business entity; neglect of measures to protect the environment; failure to make royalty payment within the time frame; non-use within one year; failure of using mineral rights (Article 25 ML).

Rights and duties

The operator has the right to choose methods of management; use the area of mineral deposits within the granted mining lease. According to Article 28 ML the operator must:

- fulfil the conditions set by the permit;
- make timely payments for subsoil use;
- make a registration of all types of operations on subsoil use;
- comply with the requirements of technical projects, plans and schemes development of appropriate mining operations to prevent excessive losses of deposit;
- comply with the conditions governing the protection of natural resources, including waste management.

4.3.1.3 Authorities

The main responsible authority for mining is the Minister of Natural Resources and Environment.

State regulation in the field of subsoil use and oversight for the safe conduct of work is done by the Minister of Natural Resources and Environment (Article 5 ML).

State control over the conduct of exploration of mineral resources has to ensure the prescribed standards (Article 49 ML).

State supervision over the management of mineral resources has to ensure that the operator uses standards (rules, standards) for safe mining operations, as well as avoids harmful influences on health and environment (Article 51 ML).

4.3.1.4 Fees and Taxes

According to Article 45 ML use of mineral resources is chargeable. Royalties are required for prospecting and exploration of minerals; payments for permits and other authorizations for the right of subsoil use; payments for mineral extraction (use of natural resources); payments for the use of geological information obtained from public funds. Royalties are collected in cash.

4.3.2 Additional Legal Basics

Table 15: List of Acts significant for raw materials – Belarus

Name of Law	No. of Law and Year of Issuing
Law on Protection of the Environment	Law No. XII of 1992 [last amendment 2009]
Law on Public Associations	1994 as amended in 2005
Law on Refugees	2003
Law on State Statistics	
Civil Code of the Republic of Belarus	
Law on payments for land	
Law on Counteraction of Monopolistic Activities and Promotion of Competition	
Law on Enterprises in the Republic of Belarus	
Law on Financial and Industrial Groups	
Law on Natural Monopolies	
Law on Income Tax	

Name of Law	No. of Law and Year of Issuing
Law on Real Estate Tax	
Tax Code of the Republic of Belarus (General part)	
Law on foreign investment	

Law on Protection of the Environment

The protection of the environment is a prerequisite of the stable economic and social development of the state. The Law is aimed at ensuring of the principles of environmental protection, nature management, protection and restoration of biological variety of natural resources and objects, and directed on provision of constitutional rights of citizens to the environment favourable to human life and health (Article 1).

5 Belgium

5.1 General Facts

The national territory covers 30,528 km² and the inhabitants are 20.665.867 (2008 estimation). The population density is 344,32/km². GDP per capita for 2009 is estimated at $ 43.794.

Constitutional Structure

The form of government is a constitutional monarchy. The political power is segregated into three levels: the Federal Government, which is based in Brussels, the three language communities and the three regions which are the Flemish Region, subdivided into 5 provinces, the Walloon Region, also subdivided into five regions and the Brussels-Capital Region.

5.2 Production of Raw Materials

Table 16: Production Data – Belgium (World Mining Data, Weber and Zsak)

Resources		2002	2003	2004	2005	2006	Change 02/06	Change 05/06
Cadmium	(t)	117	0	0	0	0	−100,00	
Baryte	(t)	29.000	28.000	27.000	30.000	0	−100,00	−100,00
Kaolin	(t)	410	429	460	460	450	9,76	−2,17
Steam-Coal	(t)	173.000	129.000	181.000	109.000	28.000	−83,82	−74,31

Table 17: Aggregates Production – Annual Statistics/Belgium, 22 April 2008, quantities in million tonnes, (UEPG 2008)

Sand & Gravel (1)		Crushed rock (2)		Marine Aggregates		Recycled Aggregates (3)		Manufactured Aggregates (4)	
2005	2006	2005	2006	2005	2006	2005	2006	2005	2006
13,9	10,1	38,0	55,5	n. a.	3,5	12,0	13,0	1,2	1,3

(1) Sand and Gravel: sold production including crushed gravel
(2) Crushed rock: sold production (excluding crushed gravel)
(3) Recycled Aggregates: materials coming from construction and demolition waste used in aggregates market.
(4) Manufactured aggregates include blast-furnace-slag, electric-arc-furnace-slag, incinerator bottom ash (IBA), pulverised fuel ash (PFA)

5.3 Normative Basics

5.3.1 Primary Legal Basics

There is no federal mining law in force anymore, as was the case prior to the creation of a Federal State in 1993. Since the Law of 8 August 1980, the management of the mineral resources has become the responsibility of the Regions (except for the continental shelf). Prior to the creation of a Federal State in 1993, the national legislation was valid for the extraction of all natural resources and a specific mining law existed. There was also a national Inspectorate of Mines, responsible for all mines and quarries. It belonged to the Ministry of Economic Affairs.[1]

5.3.1.1 General Rules

Ownership of mineral rights

Presently only industrial minerals and building materials are extracted in Belgium. As the industrial minerals and building materials are owned by the landowner, one should come to an agreement with the landowner.

5.3.1.2 Issuing of Permits

There is no regulation, specific for mineral exploration activities.

An application form has to be submitted at the community. The applicant may not be the owner of the land. After having obtained the permit, the operator should come to an agreement with the landowner. The future quarry has to be indicated on the plan for land use as an area of extraction (in fact, a pre-requisite to obtain a permit). If not, the plan for land use has to be changed first, which is a time-consuming and complex procedure. A number of criteria are used to determine an application. The standard operating conditions for mineral extraction prepared by the Wallonia Region in Belgium illustrate this point.[2] There are 52 conditions listed under the following headings:[3]

- extraction boundaries,
- use of explosives, dust control, noise control, settling ponds,
- transport on site,

---

1 Verwoort, A. (2004), Country report Belgium), in: Department of Mining and Tunnelling, University of Leoben, Minerals Policies and Supply Practices in Europe 2004.
2 Ibidem.
3 Department of Environment (2004), lc.

- protection of the aquifer-zone,
- restoration and guarantees.

The permit for the extraction of natural resources is valid for an infinite period. However, it expires if there has been no production during a two years period. The permit for the installations is, on the other hand, limited to 20 years.

Restoration

In the Order of the Walloon Government of 17 July 2003, some details are given on the rehabilitation of a surface excavation, e. g.: The rehabilitation must take into consideration the final destination of the area within the plan for land use (Article 22). The operator has to adapt the rehabilitation as a function of the geological conditions, the characteristics of the soil and the occurrence of natural plants (Article 23). The operator has to take measures for the safety of the site (e. g. stability of slopes) in the long term (Article 24). According to Articles 26–29 the cost for the rehabilitation has to be estimated prior to the start of the extraction and has to be updated yearly. The quarry has to make a bank guaranteed deposit for the entire amount, prior to the start of the extraction.[4]

5.3.1.3 Authorities

A separate Inspectorate of Mines does not do the inspection in the Regions anymore. The extraction industry is considered in a similar way, as other industrial activities. Most of the monitoring and the enforcement are done on the level of the Region. For example the safety aspects are the responsibility of the Ministries of Employment, while the environmental aspects are for the Ministries of Environment. The situation for the use of explosives is more complex: (1) the federal Ministry of Economic Affairs is involved in providing a permit and in the control of it, (2) the regional Ministries of Environment are involved for some environmental aspects, and (3) the Provincial Governor gets involved when an accident occurs.

5.3.1.4 Fees and Taxes

Administration fees are payable after having received a permit. Industrial activities in the Walloon Region need an environmental permit, if one has already a landscape permit, or a so-called unique environmental permit, that includes the landscape permit. The fees are 125 Euro and 500 Euro respectively.[5]

---

[4] Verwoort, A. (2004), lc.
[5] Ibidem.

The extraction of gravel in the Province of Limburg is a special case. It is regulated by the Gravel Decree of 14 July 1993, aiming at a systematic reduction of river gravel production in the Belgian Province of Limburg. A fee is charged per tonnage gravel extracted and a foundation has been created to manage this money. The aim of the foundation is among others to study alternatives for gravel, to help in the re-structuring of this branch of the industry, to assist with the social aspects for the people working in the quarries, etc.

5.3.2 Additional Legal Basics

Table 18: List of Acts significant for raw materials – Belgium

Region	Name of Law	Date of Issuing
Continental shelf (Federal level)	Law of 13 June 1969 amended by the Law of 20 January 1999 and the Law of 22 April 1999 on exploration and exploitation of sand and gravel on the continental shelf.	1968 last amended in 1999
Flemish Region	Decree on Surface Mineral Resources ('Oppervlaktedelfstoffen Decreet') of 4 April 2003, followed by the Order of the Flemish Government of 26 March 2004.	2003 last amended 2004
	Flemish Decree of 18 May 1999 with regard to town and country planning.	1999
	Decree of 28 June 1985 with regard to environmental permits.	1985
	Decree of 23 March 1989 on the Environmental Impact Evaluation.	1989
	Decree of 28 June 1985, defining the Flemish regulation on environmental permits (Vlarem I and II).	1985
	Decree of 2 July 1981 with regard to the management of waste.	1981
	Decree of 12 December 1990 on the authority of OVAM.	
	Decree of 22 February 1995, Decree of 26 May 1998 and Order of 5 March 1996, linked to VLAREBO regulation.	
	VLAREA regulation, approved on 17 December 1997.	
	Gravel Decree of 14 July 1993.	

Region	Name of Law	Date of Issuing
Walloon Region	SDER: Development scheme of the regional space ("Schéma de Développement de l'Espace Régional"), accepted by the Walloon Government on 27 May 1999.	1999
	PEDD: Environmental plan for sustainable development ("Plan d'Environnement pour le Développement Durable"), accepted by the Walloon Government on 9 March 1995.	1995
	CAWA: Contract for the future of the Walloon Region ("Contrat d'Avenir pour la Wallonie"), developed by the Walloon Government since 1999.	1999
	DPR: Declaration of regional politics ("Déclaration de Politique Régionale") from 15 July 1999.	1999
	Decree of 18 December 2003, modifying the Decree of 11 March 1999, related to the environmental permit.	1999 last amended in 2003
	Order of the Walloon Government of 4 July 2002 (published on 1 October 2002) determining the general conditions for exploitations.	2002
	Order of the Walloon Government of 4 July 2002 (published on 21 September 2002) organising the evaluation of environmental incidences in the Walloon Region.	2002
	Order of the Walloon Government of 4 July 2002 determining the list of projects subjected to an incidence study and of classified installations and activities.	2002
	Order of the Walloon Government of 4 July 2002 (published on 21 September 2002), related to the procedures and the various measures to apply the Decree of 11 March 1999, related to the environmental permit.	1999 last amended in 2002
	Order of the Walloon Government of 17 July 2003 (published on 6 October 2003), on the "sector conditions" for quarries and their installations.	2003

6 Bulgaria

6.1 General Facts

The national territory covers 110.912 km² and the inhabitants are about 7,6 million. The population density is about 69/km². The GDP/per capita amounts to $ 6.223 (2009).

Constitutional Structure

The form of government is a parliamentary republic. The structure of the state consists of 28 districts (oblast). Each of the districts has its own governor.

6.2 Production of Raw Materials

Table 19: Production Data – Bulgaria (World Mining Data, Weber and Zsak, 2008)

Resources		2002	2003	2004	2005	2006	Change 02/06	Change 05/06
Iron	(t)	217.553	205.000	200.000	198.000	100.000	−54,03	−49,49
Manganese	(t)	11.700	12.000	12.300	11.358	6.048	−48,31	−46,75
Cadmium	(t)	345	307	356	319	320	−7,25	0,31
Copper	(t)	84.400	91.600	79.600	94.900	84.000	−0,47	−11,49
Lead	(t)	19.800	20.000	19.000	13.000	13.000	−34,34	0,00
Zinc	(t)	14.000	12.000	11.000	11.956	6.000	−57,14	−49,82
Gold	(kg)	2.612	2.270	2.431	3.868	3.818	46,17	−1,29
Silver	(kg)	60.000	60.000	60.000	60.000	50.000	−16,67	−16.67
Baryte	(t)	80.000	50.000	20.000	10.000	0	−100,00	−100,00
Gypsum	(t)	156.000	154.000	152.000	150.000	185.000	18,59	23,33
Kaolin	(t)	100.000	150.000	200.000	235.062	282.260	182,26	20,08
Salt	(t)	1.294.200	1.353.158	1.358.400	1.366.100	1.438.000	11,11	5,26
Sulphur	(t)	19.923	20.300	20.500	20.700	21.400	7,41	3,38
Steam Coal	(t)	14.500	14.800	15.000	17.000	51.900	257,93	205,29
Brown Coal	(t)	23.201.600	23.000.000	26.400.000	22.205.000	22.749.500	−1,95	2,45
Natural Gas	(Mm³)	11	10	11	10	20	81,82	100,00
Oil	(t)	33.000	35.000	37.000	30.010	25.300	−23,33	−15,69

Table 20: Production of aggregates – Bulgaria. (Metric tons unless otherwise specified), (USGS, 2008)

Bulgaria		2002	2003	2004	2005	2006
Limestone and dolomite	1000 metric tons	3.500	3.500	3.500	3.265	3.340
Sand and gravel	1000 m³	2.385	2.098	3.333	3.628	4.293
Silica, quartz sand	1000 metric tons	607	412	545	583	250

6.3 Normative Basics

6.3.1 Primary Legal Basics

The primary legal basics of mineral extraction activity is the Mining Law ("Law on Mineral Resources") No. 23 of 1999. Besides that there exist the Mines and Quarries Act No. 92 of 1957 as amended by Law No. 35 of 1996.

6.3.1.1 General Rules

Table 21: Structure of the Mining Law 1999 – Bulgaria

Part 1 – Common Provisions
Chapter 1: Subject – Matter and Scope Articles 1–5
Chapter 2: Mineral Resources Management Authorities Articles 6–7
Chapter 3: Public Procurement of Geological Surveys – Articles 8–11
Chapter 4: Geological and Technical Information
Chapter 5: National Geofund – Article 16
Chapter 6: Registers and Cadastres of the Permits for Prospecting and/or Exploration and the Extraction Concessions – Articles 17–19
Chapter 7: National Balance of Reserves and Evaluation of Resources. Register of Discoveries and Specialized Cadastre of Deposits – Articles 20–22
Chapter 8: Mining Waste Management Articles 22a–22k

6.3 Normative Basics

Part 2 – Prospecting for and Exploration and Extraction of Mineral Resources

Chapter 1:	Permits for Prospecting and/or Exploration and Extraction Concessions	Section I: Common Provisions Articles 23–26 Section II: Permits for Prospecting and/or Exploration Articles 27–32 Section III: Extraction Concession Articles 33–38
Chapter 2:	Terms and Conditions for Granting Permits for Prospecting and/or Exploration and Extraction Concessions	Section I: Common Provisions Articles 39–41 Section II: Granting of Permits for Prospecting and/or Exploration and Extraction Concessions through Competitive Bidding or Tender Articles 42–50 Section III: Direct Granting of Permits for Prospecting and/or Exploration and Extraction Concessions Articles 51–57
Chapter 3:	Financial Terms and Conditions	Articles 58–64
Chapter 4:	Contracts, Termination; Arbitration and Experts Opinion	Section I: Common Provisions Articles 65–66 Section II: Termination Articles 67–71 Section III: Arbitration and Expert's Opinions Articles 72–73
Chapter 5:	Use of Land	Articles 74–76

Part 3 – Protection of the Earth Recesses and Reasonable Use of Mineral Resources

Chapter 1 Common Provisions Articles 77–79

Chapter 2: Mines and Quarries Articles 80–81

Chapter 3: Work Plans Articles 82–86

Chapter 4: Protection of the Earth Recesses Articles 87–89

Chapter 5: Enforcement of the Protection of the Earth Recesses and the Reasonable Use of Mineral Resources Articles 90–92

Chapter 6: Administrative Penalties Articles 93–97

Additional Provisions

Transitional and Concluding Provisions

According to Article 2 ML, mineral resources are grouped into:
1. metallic mineral resources;
2. non-metallic mineral resources – industrial minerals;
3. oil and gas;
4. solid fuels;
5. building materials;
6. stone cutting materials;
7. mining waste

Ownership of mineral rights

According to Article 3 ML mineral resources are exclusive state property.

Acquiring mineral rights

Article 4 ML states that prospecting for or exploration of mineral resources is based on a permit. Mineral resources must be extracted on the basis of a concession. According to Article 23 (1) ML licenses for prospecting and/or exploration and extraction concessions are granted to natural persons and legal entities, provided that they are duly registered as traders and have the technical, managerial and financial capabilities required for the performance of the respective activities.

More than one permit for prospecting and/or exploration and extraction concession may be granted for the same area, provided that it is granted for different types of mineral resources and the activities associated with one permit will not interfere with the activities associated with any other permit (Article 23 (2) ML).

According to Article 26 ML proceedings in relation to granting a license for prospecting and/or exploration and for a concession can be opened after coordination with the competent Ministries for Protection of the National Security and Defence of the country when protected areas, sites and cultural values are concerned.

6.3.1.2 Issuing of Permits

Prospection and exploration

Article 28 ML regulates the rights and duties of a holder of a prospecting and exploration permit. The holder can within the granted area perform all necessary activities aimed at discovering and evaluating deposits of mineral resources, for which the permit is issued; evaluate deposits of mineral resources, for which the permit is issued, including the extraction for technological testing; submit a statement in pursuance of Article 21 ML for its declaration as a commercial discovery in order to be duly registered.

The holder of a license for prospecting and/or exploration must provide the information under Art. 13 ML to the National Geofund (Article 30 ML). According to Article 31 ML the validity term of a license for prospecting and/or exploration is up to three years and may be renewed twice by up to two years. Article 32 (1) ML describes the allowed size of prospecting/exploration area for different mineral resources.

Article 39 (1) ML regulates the conditions concerning permits for prospecting for and/or exploration of mineral resources. Mineral resources covered by Article 2 ML can be granted through: 1. Competitive bidding; 2. Tender; 3. direct selection of the permit holder, where the latter is the only applicant upon the expiration of a one-month period after the publication of the announcement for upcoming permit granting in the State Gazette and the Internet site of the competent body under Article 7 ML.

Exploitation

According to Article 39 (2) ML concessions for extraction of mineral resources can be granted either through competitive bidding or tender; or to a holder of a permit for prospecting and exploration or a license for exploration by right in pursuance of Art. 29 ML; by right to a company on the grounds of a privatization deal.

According to Article 29 ML the holder of a permit for prospecting and exploration or for exploration can directly be selected as a concessionaire for extraction of a discovered deposit upon the fulfilment of the following conditions: He has stated and registered in pursuance of Article 21 (3) the discovery of a deposit of mineral resources within the validity term and the area of the permit; he has obtained a commercial discovery certificate in pursuance of Article 21 (7).

According to Article 33 ML extraction concessions can be granted for specific deposits of mineral resources or parts (sections) thereof. The extraction concession entitles the concessionaire to perform all necessary activities related to the extraction, and sale of the mineral resources and the use of mining waste, for which the concession is granted (Article 34 ML).

The validity term of the extraction concession is up to 35 years and can be renewed by up to 15 years (Article 36 ML). Article 37 ML determines the concession area. The concession area shall cover the area including the deposit and the areas needed for the performance of the activities. Articles 51–57 ML regulate the (directly) issuing of permits and concessions. The issuing of permits and concessions is based on a competitive bidding or tender according to Article 42–50 ML.

Rational utilization of resources

Article 80 states that the boundaries of each mine or quarry have to be established in accordance with the outlines the deposit and resources which are the subject of exploitation. Additionally according to Article 81 certain documents have to be prepared for each mine and quarry including precise topographic map of the mine or quarry field in an appropriate scale, indicating the boundaries, cartographic signs of the location of the mine or quarry, the nature of the locality and all installations and equipment thereon; the required registers, maps and statistical data related to the operational requirements and the safety and health of employees in mines and quarries.

Articles 82–86 ML include regulations concerning the establishment of operating (exploitation) plans.

According to Article 82 ML the prospecting for, exploration, extraction and primary processing of mineral resources, the technical liquidation and conservation of geological survey and mining facilities have to be performed on the basis of general and annual work plans drawn up by the holders of licenses for prospecting and/or exploration or the concessionaires after coordination with the competent authorities under Article 7 (1). The general and annual plans for prospecting, exploration, extraction and primary processing of mineral resources have to be subject to mandatory coordination with the Minister of Environment and Waters concerning the requirements to bowels of the earth conservation, the rational use of subsurface resources and the measures for conservation and reclamation of the environment. (2).

Article 86 obliges the holders of licenses for prospecting and/or exploration or concessionaires to report annually the implementation of the work plans under Articles 83 and 84 through a written report to the Minister with whom the contract has been concluded, as well as to the Minister of Environment and Waters.

6.3.1.3 Authorities[1]

Competences of authorities are mentioned in Article 5 ML and Article 6–7 ML. According to Article 5 mineral rights are granted through licenses for prospecting and/or exploration, concession for exploitation issued by the respective authorities in accordance with their powers under Article 7 ML.

Article 7 ML is registering different authorities. These are amongst other causes the Minister of Environment and Water, the Minister of Economy, Energy and Tourism, the Minister of Regional Development and Public Works. All these authorities have clearly structured functions. The Minister of Environment and Water issues permits regarding registered deposits, issues in line with the bidding procedure prospecting and exploration licences related to mineral categories 1, 2, 4 to 7 (Article 7 (1) ML).

According to Article 7 ML the Minister of the Environment and Waters is responsible for the prospecting for, exploration and extraction of mineral resources (1). The Minister of Economy, Energy and Tourism has to hold competitive bidding and tenders, conduct negotiations and issue licenses for the prospecting for and/or exploration of mineral resources under Article 2, categories 1, 2, 4 and 7. He also has to provide the National Geofund with the geological and technical information collected under Art. 13 by the license holders and concessionaires (2). The Minister of Regional Development and Public Works is responsible for the extraction of construction minerals (3).

According to Article 90 (1) ML the Ministry of the Environment and Waters has to supervise the enforcement of the protection of the earth recesses the reasonable use of mineral resources, as well as the control over the implementation of the plans for conservation, liquidation, and recultivation. In cases when objects of cultural value are concerned, the Ministry of Culture is responsible for the control. According to Article 91 ML the enforcing authorities for the protection of the earth recesses and the reasonable use of mineral resources are entitled to unhindered access to all facilities, buildings and equipment of the holder of the permit for prospecting and/or exploration or the concessionaire within the boundaries of the granted area.

---

1 It should be considered that with the last amendments of the Mining Law (coming into force at the end of November 2010) there is only one competent authority (Unified Body) which will be a structure of the Ministry of Economy, Energy and Tourism.

6.3.1.4 Fees and Taxes

According to Article 58 ML the rights to prospect for and/or explore or to extract mineral resources through a permit for prospecting and/or exploration or through an extraction concession are granted against payment.

According to Article 60 (1) ML the holder of the permit for prospecting and/or exploration must pay an annual fee for the area granted. The amount of the fee is determined in accordance with validity term of the permit, the size of the area granted and the group of mineral resources, for which the permits are granted (2).

According to Article 61 (1) ML the concessionaire must pay a concession fee. According to Article 61 (6) for deposits of mineral resources with unfavourable mining and geological, technological and economic characteristics, the concessionaire may be temporarily exempted from payment of the concession fee for a period of 5 years or the concession fee may be reduced by up to 50 per cent of the agreed fee for a period of up to 5 years.

6.3.2 Additional Legal Basics

Table 22: List of acts significant for raw materials – Bulgaria

Name of Law	No. of Law and Year of Issuing
Environment Protection Act	Law No. 86 of 1991, as amended by Law No 96 of 2002
Nature Protection Act	Law No. 47 of 1967, as amended by Law No. 86 of 1991
Clean Air Act	1996
Law of the Purity of Atmospheric Air	1996
Water Act	Law No. 67 of 1999
Waste Management Act	Law No. of 2003, as amended in 2007
Concessions Act	Law No. of 1995, as amended in 1999
Administrative Procedure Act	
Concessions Act	
Law on Local Government and Local administration	
Law on National Standardization	
Law on the Statistics	
Public Procurement Act	
Civil Procedure Code	

Name of Law	No. of Law and Year of Issuing
Obligations and Contracts Act	
Ownership Act	
State Property Act	
Act on the Special Investment Purpose Companies	
Cooperatives Act	
Tax Administration Act	
Tax on Profits Act	
Export Insurance Act	
Insurance Code	

Environmental Protection Act

This Act regulates the protection of the environment for the present and future generations and protection of human health (Article 1). Conservation and use of the bowels of the Earth upon prospecting, exploration and extraction of subsoil resources, including use of past mining and processing work sites, shall follow a procedure established by this Act and by the Subsoil Resources Act (i. e. mining law) (Article 48).

Water Act

This Act regulates the ownership and management of waters within the territory of the Republic of Bulgaria as a national indivisible natural resource and the ownership of the water systems and facilities (Article 1). For performance of drilling and/or mining operations in areas possessing substantial ground water resources as designated by an order of the Minister of Environment and Water, consultation with the Ministry of Environment and Water are required in respect of the terms and conditions for use of the water bodies containing ground waters (Article 51).

7 Croatia

7.1 General Facts

The national territory covers 56 542 km² and the inhabitants are 4.489.409 (2009 estimation). The population density is 81/km². GDP per capita for 2009 is estimated at $ 15.283.

Constitutional Structure

The form of government is a parliamentary republic. It is divided into 20 counties and the capital city of Zagreb.

7.2 Production of Raw Materials

Table 23: Production Data – Croatia (World Mining Data, Weber and Zsak, 2008)

Resources		2002	2003	2004	2005	2006	Change 02/06	Change 05/06
Bauxite	(t)	26.521	15.000	1.482	500	600	−97,74	20,00
Bentonite	(t)	1.000	900	671	1.472	1.200	20,00	−18,48
Gypsum	(t)	206.505	150.000	77.987	207.918	297.894	44,26	43,27
Salt	(t)	19.000	18.000	16.650	17.100	18.496	−2,65	8,16
Lignite	(t)	0	0	0	0	80.000		
Nat. Gas	(Mm³)	2.880	2.700	2.352	2.432	2.864	−0,56	17,76
Oil	(t)	906.953	900.000	802.615	745.589	728.651	−19,66	−2,27

Table 24: Aggregates Production – Annual Statistics/Croatia, 22 April 2008, quantities in million tonnes (UEPG 2008)

Sand & Gravel (1)		Crushed Rock (2)		Marine Aggregates		Recycled Aggregates (3)		Manufactured Aggregates (4)	
2005	2006	2005	2006	2005	2006	2005	2006	2005	2006
n.a.	6,2	n.a.	21,8	n.a.	0,0	n.a	0,0	n.a.	0,0

(1) Sand and Gravel: sold production including crushed gravel
(2) Crushed rock: sold production (excluding crushed gravel)

(3) Recycled Aggregates: materials coming from construction and demolition waste used in aggregates market
(4) Manufactured aggregates include blast-furnace-slag, electric-arc-furnace-slag, incinerator bottom ash (IBA), pulverised fuel ash (PFA)

7.3 Normative Basics

7.3.1 Primary Legal Basics

The primary legal basics of mineral extraction activity is Mining Law No. 75 of 2009.

7.3.1.1 General Rules

Table 25: Structure of the Mining Law – Croatia

I.	General Provisions	Articles 1–18
II.	Exploration of Mineral Raw Materials	Articles 19–35
III.	Mineral Raw Materials Reserves	Articles 36–40
IV.	Mining Projects	Articles 41–44
V.	Exploitation of Mineral Raw Materials	Articles 45–77
VI.	Construction of Mining Facilities and Installations	Articles 78–108
VII.	Mining Plans and Measurements	Articles 109–112
VIII.	Single Information System of Mineral Raw Materials	Articles 113–114
IX.	Qualifications required for the Performance of specific Activities	Articles 115–117
X.	Occupational Health and Safety Measures	Articles 118–124
XI.	Administrative and Inspectional Supervision	Articles 125–127
XII.	Penal Provisions	Articles 128–131
XIII.	Transitional and Final Provisions	Articles 132–140

The mineral raw materials are classified as follows (Article 5 ML):

1. energy resources – all sorts of fossil coal, carbohydrates in solid, liquid or gaseous state, all sorts of bituminous and oil rocks, other gases found inside the earth and radioactive mineral raw materials;
2. mineral raw materials out of which metals and their compounds could be produced;
3. non-metallic mineral raw materials – graphite, sulphur, magnesite, fluorite, barite, asbestos, mica, phosphate, gypsum, calcite, chalk, bentonite clay, flint stone, flint sand, kaolin, ceramic and refractory clay, feldspar,

talc, tuff, raw materials for the production of cement and lime, carbonate and silicate raw materials for industrial processing;
4. dimensional stone;
5. all kinds of salts and saline waters;
6. mineral and geothermal waters out of which mineral raw materials can be extracted;
7. artificial building stone, construction sand, gravel and brick clay.

Ownership of mineral rights

Mineral resources are in the ownership of the Republic of Croatia (Article 3 ML).

Acquiring mineral rights

The operator can acquire the right to explore and exploit mineral raw materials within a certain territory on the basis of a mining concession granted by a government authority in charge of mining activities. The mining concession is based on the concession contract concluded between the government authority and the applicant. The concession for mineral raw materials exploitation gives the right to perform mining works with the aim to use mineral raw materials for economic purposes (Article 46 ML).

7.3.1.2 Issuing of Permits

The permit for the exploration of mineral raw materials or mining concession for exploitation of mineral raw materials will be granted to a physical person or legal person with headquarters in the Republic of Croatia registered for these activities (Article 12 ML).

The mining concession is not transferable to another physical or legal person without consent of the government authority that issued them (Article 14 ML).

The operator is obliged to exploit the mineral raw materials deposit rationally (Article 63 ML).

Exploration

An approval for exploration of mineral raw materials shall be issued on the basis of a public tender.

The administrative decree by which the exploration is approved has to prescribe an appropriate design, i. e. technical documentation to be elaborated and submitted for verification according to the provisions of this Law (Article 22 ML).

The exploration of mineral raw materials will be permitted only within the area approved for exploration. The data on the approved exploration areas shall be entered into the Cadastre of Exploration Areas (Article 19 ML).

The exploration of mineral raw materials is basically not allowed within the area where urban settlements, public traffic routes, water supply facilities and equipment, military objects, monuments, especially protected natural objects and cemeteries are located, as well as within the area where other facilities of public importance determined by special regulations are situated (Article 23 ML).

The operator is obliged to report on the beginning of exploration works to the State Inspectorate and to the county (regional) office in charge of mining activities on whose territory minerals shall be explored 15 days prior to the beginning of works at the latest.

During the exploration, mineral raw materials can be extracted for laboratory investigation, technological tests, and for the establishment of conditions for exploitation in quantities stipulated in the exploration permit (Article 30 ML). During the exploration works the operator is obliged to submit reports on the exploration works performed and on the results achieved to the government authority that issued the permit (Article 26 ML).

After the termination of the exploration works, the operator is obliged to undertake all the security measures that exclude the possibility of dangerous events for people, property and environment on the locations where the exploration works had been performed and to inform thereupon the county (regional) office in charge of mining activities (Article 31 ML).

Exploitation

For the exploitation permit the following is needed (Article 45 ML):

1. permit for exploitation field,
2. a mining concession for mining works,
3. a building permit for the construction of mining facilities and installations,
4. a permit to use of mining facilities and installations with supplied with electronic weightbridge as prescribed.

The exploitation of mineral raw materials is permitted only within the approved exploitation field. Data on the approved exploitation field approved must be entered into the Exploitation Fields Cadastre (Article 32 ML).

If a request for the exploitation field permit is submitted by the operator for a site where he had previously performed the exploration, the permit can be granted to the operator that meets the prescribed exploitation conditions. If the operator does not fulfil the prescribed exploitation conditions, the operator receiving the exploitation field is obliged to refund the means invested into the exploration works to the operator who performed the mineral raw materials exploration works (Article 32 ML).

Mining projects can be elaborated only by a legal person registered for the elaboration of mining projects (Article 42 ML).

The general and additional projects are subject to a special verification regarding the solutions on rational exploitation of mineral raw materials, measures and standards of industrial protection, on security of facilities and people, as well as the security of underground, surface and neighbouring objects (Article 43 ML).

In granting a mining concession for mining works the competent authority has to consider (Article 53 ML):

– the applicant's capacity to realise the concession,
– the financial benefits from the realisation of the concession,
– the impact on environmental protection and preservation.

The period for which the mining concession is granted is determined based on the quantity and diffusion of the established remaining reserves of the mineral material in the exploitation field, which enables the operator to exploit the field for a maximum duration of 40 years, with a projected annual production (Article 54 ML).

The mining concession for the mining works must include (Article 58 ML):

1. location permit for intervention in the exploitation field;
2. mining design for mineral raw materials exploitation;
3. deadline within which the Concessionaire must commence to realise the concession;
4. name of institutions and companies which must be notified of the concession realisation commencement for mineral raw materials exploitation.

The operator is obliged to report the beginning of the realisation of the mining concession for the mining works, i. e., the beginning of construction of mining facilities and installations to the State Inspector's Office, ministry competent for mining, ministry competent for finances and the central state administration body competent for state property management, at least 15 days prior to the beginning of mining works, and construction works respectively (Article 67 ML).

If the work performed in pits and open-cast mines is temporarily interrupted due to unexpected events, the operator is obliged to inform the State Inspectorate regarding these circumstances within 24 hours after the work stoppage (Article 68 ML).

The Minister of Economy may, after acquiring a consent from the Minister of Environmental Protection and Urban Planning, enact a more detailed regulation regarding the content of request for the exploitation field, public debate procedure regarding the approval of the exploitation field and the manner of determining its size (Article 48 ML).

Extraction plan

The operator exploring or exploiting mineral raw materials is obliged, on the basis of measurements, to implement and supplement the extraction plans (Article 110 ML).

The operator exploring and exploiting mineral raw materials must have: 1. the site plan of the exploration or exploitation field; 2. Cadastre Plan Certificate with delineated boundaries of exploration area i. e., exploitation field, 3. the geologic map of the exploration area, i. e. geologic maps of the exploitation field and of its surroundings, as well as the characteristic geologic profiles (Article 111 ML).

Cadastre on approved exploitation fields

The competent authority that issues the exploration and exploitation permit, is obliged to provide single information system for mineral raw materials, comprising of the Register of approved mineral raw materials exploration areas and the Register of approved mineral raw materials exploitation fields. Besides the Cadastre the government authority issuing the permit is obliged to keep records on all exploration sites demanded, and to establish a collection of documents and a list of trading companies and craftsmen that were granted the permit for exploitation field, or, granted mining concession for mining works (Article 113 ML).

7.3.1.3 Authorities

The main responsible authority for mining is the Minister of Economics.

According to Article 21 ML the permit for the exploration of mineral raw materials under Article 5 will be granted by the Ministry in charge of mining.

According to Article 47 ML the minister in charge of mining grants the concession for exploitation of mineral materials referred to in Article 5.

Government supervision of the implementation of this Law and regulations based thereupon is performed by the Ministry in charge of mining. The inspection is implemented by the State Inspectorate in accordance with a special law (Article 125 ML).

7.3.1.4 Fees and Taxes

The operator has to pay a fee for the exploitation of the mineral raw materials (Article 28 ML).

Fees and taxes are regulated by the Ordinance on fee for exploration of mineral raw materials (Official Gazette No. 158/2009) and by the Ordinance on fee for the concession for exploitation of mineral raw materials (Official Gazette No. 158/2009).

7.3.2 Additional Legal Basics

Table 26: List of acts significant for raw materials – Croatia

Name of Law	No. of Law and Year of Issuing
Environmental Protection Act	Law No. 110 of 2007
Air Protection Act	Law No. 178 of 2004, amended by Law No 60 of 2008
Water Management Act	Law No. 153 of 2009
Nature Protection Act	Law No. 70 of 2005, amended by Law No 149 of 2008
Waste Act	Law No 178 of 2004, last amended by Law No 87 of 2009
Act on National Statistics	Law No. 103 of 2003, amended by Law No 75 of 2009
Public Procurement Act	Law No. 119 of 2007, amended by Law No 125 in 2008
Civil Obligations Act	Law No. 35 of 2005, amended by Law No 41 in 2008

Name of Law	No. of Law and Year of Issuing
Civil Procedure Act	Law No. 53 of 1991, last amended by Law No. 123 of 2008
Crafts Act	Law No. 49 of 2003, last amended by Law No. 79 of 2007
Land Registration Act	Law No. 91 of 1996, last amended by Law No. 152 of 2008
Act on Foundations and Funds	Law No. 64 of 2001
Competition Act	Law No. 79 of 2009
Labour Act	Law No. 149 of 2009
General Tax Act	Law No. 147 of 2008
Real Estate Sales Tax Act	Law No. 53 of 1990, last amended by Law No. 153 of 2002
Ordinance on the Methods and Conditions for the Landfill of Waste, Categories and Operational Requirements for Waste Landfills	Law No. 117 of 2007
Regulation on Limit Values of Pollutants in Air	Law No. 133 of 2005
Concession Act	Law No. 125 of 2008

Environmental Protection Act

Environmental Protection Act regulates (Article 1): environmental protection and sustainable development principles, protection of environmental components and protection against environmental burdening, actors in environmental protection, sustainable development and environmental protection documents, environmental protection instruments, environmental monitoring, information system, ensuring access to environmental information, public participation in environmental matters, access to justice, liability for damage, financing and instruments of general environmental policy, administrative and inspection supervision. According to Article 56 mining projects require a mandatory strategic assessment.

Water Management Act

This Act regulates the legal status of water and water estate, the methods and conditions of water management (water use and protection, regulation of watercourses and other water bodies, and protection from adverse effects of water), the method of organizing and performing water management tasks and functions, basic conditions for carrying out of water management activities (Article 1). Sand and gravel in renewable deposits in areas of importance to the water regime may be excavated only on the basis of the water rights permit and of the concession contract. The areas of importance to the water regime pursuant to Article 53 of the Water Management Act are watercourses and other surface water bodies (as described in Article 85). For the excavation of sand and gravel from non-renewable deposits and for the excavation of rock in the area of watercourses and other surface water bodies, and in the sanitary protection zone, in addition to conditions determined by the Act on Mining, a water rights permit is required (Article 53).

Nature Protection Act

The purpose of the Nature Protection Act is to regulate the system of protection and integrated conservation of nature and its values. The term nature includes the overall biological and landscape diversity (Article 1). Article 28 of the Nature Protection Act regulates mining issues as follows:

Exploration and exploitation of mineral resources must be carried out in such a manner as to conserve landscape values of the space to the highest degree possible (1). Harmful effects on a landscape likely to be caused by exploitation of mineral resources must be avoided by selecting the most favourable site, type and scope of the activity planned (2). Harmful effects on a landscape caused by exploration and exploitation of mineral resources must be eliminated by restoration of the extraction site or by the arrangement of the entire exploitation field (3). The statement about the examination of a mining project and approval of the project design may not be issued unless the Ministry issues a certificate confirming that the project design of remediation or arrangement of the entire exploitation field has been prepared in conformity with the nature protection conditions prescribed (4). Remediation of the exploitation site affected by the activities upon completion of exploitation of mineral resources form a constituent part of the main mining project for mineral resources exploitation (5).

8 Cyprus

8.1 General Facts

The national territory covers 9,251 km² and the inhabitants are 793.963 (2009 estimation). The population density is 117/km². GDP per capita for 2009 is estimated at $ 29.619.

Constitutional Structure

The form of government is a presidential republic. It is divided into 6 districts (Nicosia, Famagusta, Kyrenia, Larnaca, Limassol and Paphos)

8.2 Production of Raw Materials

Table 27: Production Data – Cyprus (World Mining Data, Weber and Zsak, 2008)

Resources		2002	2003	2004	2005	2006	Change 02/06	Change 05/06
Copper	(t)	3.695	2.552	1.334	0	900	−75,64	
Bentonite	(t)	128.000	144.859	155.717	172.366	150.620	17,67	−12,62
Gypsum	(t)	182.000	191.000	255.000	210.000	264.000	45,05	25,71

Table 28: Production of aggregates – Cyprus (Metric tons) (USGS, 2008)

	2002	2003	2004	2005	2006
Sand and stone:					
Limestone, crushed (Havara)	2.000	1.000	1.200	1.000	700
Marble	2	2	1	2	1
Marl, for cement production	1.950	2.220	2.290	2.450	2.210
Sand and gravel	10.500	10.700	11.600	12.064	12.199
Building stone	80	103	105	51	57

8.3 Normative Basics

8.3.1 Primary Legal Basics

The primary legal basics of mineral extraction activity is the Mining Law ("Mines and Quarries Law") No. 5 of 1965, as amended by Law No. 63 of 2003.

8.3.1.1 General Rules

Table 29: Structure of the Mining Law – Cyprus

Part I Preliminary	Section 1: Short title	
	Section 2: Interpretation	
	Section 3: Meaning of "mine" and "quarry"	Item 1–5
Part II General Provisions	Section 4: Ownership of minerals and quarry materials	Item 1–2
	Section 5: Prohibition of prospecting, mining or quarrying	
	Section 6: Law does not apply to mineral oils etc.	
	Section 7: Holder of prospecting permit etc. not resident in the Colony	Item 1–3
	Section 8: Plans of prospecting, mining and quarrying operations	
	Section 9: Capital	Item 1–3
	Section 10: Lands excluded from prospecting, mining and quarrying	
	Section 11: Royalties	Item 1–2
Part III Prospecting	Section 12: Prospecting when lawful	Item 1–5
	Section 13: Prospecting permits	Item 1–4
	Section 14: Rights of the holder	Item 1–2
	Section 15: Duties of the holder	Item 1–2
	Section 16: Ownership and disposal of minerals or quarry materials raised	Item 1–4
	Section 17: Payment of compensation by the holder of a permit	Item 1–6
	Section 18: Cancellation of prospecting permit	

	Section 19: No second prospecting permit for the same area	
	Section 20: Priority	Item 1–2
	Section 21: Governor may close areas for prospecting	Item 1–3
	Section 21A: A Survey, geological prospecting, etc.	Item 1–3
Part IV Mining	Section 22: When mining lawful	Item 1–2
	Section 23: Grant of a mining lease	Item 1–5
	Section 24: Duration, renewal and surrender of mining lease	Item 1–7
	Section 25: Implied conditions on part of lessee	
	Section 26: Rights under a mining lease	Item 1–10
	Section 27: Transfer or assignment of a mining lease	
	Section 28: Diversion or pollution of waters prohibited	Item 1–3
	Section 29: Purification of water	
	Section 30: Power to Inspector of Mines to enter upon land and inspect etc.	
	Section 31: Power to Governor to close mine	Item 1–2
	Section 32: Power to Governor to determine mining lease	
	Section 33: Mine plans etc. to be lodged on abandonment or closing down of mine	Item 1–2
	Section 34: Right of lessee to remove fixtures	Item 1–2
	Section 35: Acceptance of rent not to act as a waiver of forfeiture	

Part V Quarrying	Section 36: Control of quarries in Governor	
	Section 37: Quarrying lawful under permit or licence	Item 1–2
	Section 38: Period, renewal and revocation of quarry permits	Item 1–2
	Section 39: Covenants etc. for quarry licences	Item 1–3
	Section 40: Power to impact or close quarries	Item 1–3
	Section 41: Power to Governor to determine quarry permit	
Part VI Miscellaneous	Section 42: Appointments and duties of mines officers	Item 1–2
	Section 43: Offences and penalties	Item 1–2
	Section 44: Liability of employer for offences committed by servant	
	Section 45: Bounding of prospecting permits, etc.	
	Section 46: Materials required for public works	
	Section 47: Power to Governor in Council to make Regulations	
	Section 48: Saving	

Ownership of minerals

According to Article 4 (1) ML the ownership in, and control of all minerals and quarry materials belongs to the State.

Acquiring mineral rights

According to Article 26 (1) ML a mining lease confers the right to carry out mining operations (from natural or legal persons) below the surface of the ground within the area for which such lease has been granted.

8.3.1.2 Issuing of Permits

Exploration

According to Article 13 (1) ML the Governor can grant to any person an exploration permit based on payment of the prescribed fees. Such a permit is not transferable (3). An exploration permit remains in force for one year but can be renewed by the Governor.

According to Article 14 (1) ML the holder of an exploration permit has the right to enter upon and explore any state's owned land. However, he has no right to explore on private land unless he obtains the consent of its owner.

The holder of an exploration permit must carry on all exploration activities in a safe and workmanlike manner. He must keep registers and permit at all reasonable times the Inspector of Mines to inspect any activities (Article 15 ML).

Exploitation

The Governor may grant a mining lease to any person applying in the prescribed manner if the mineral-bearing qualities and quantities of the land in the area applied for are such as to justify the grant and that the applicant possesses sufficient working capital and technical knowledge to ensure the proper development and exploitation of the area applied for (Article 23 ML). A mining lease must specify the area and the minerals in respect of which it is granted and is subject to the payment of fees, royalties (2). The Governor in granting a mining lease can require restoration of any area used for the mining operations (3).

A mining lease can be granted for not more than 50 years (Article 24 ML).

According to Article 25 ML every mining lease implies certain condition, for instance the operator must

- pay rent and any royalty that may be required;
- commence mining within a period of six months from the date of the lease;
- carry on all mining work in a safe and workmanlike manner;
- provide extraction plans and books of account, e. g. for inspection.

Part V – Quarrying

According to Article 38 (1) ML a quarry permit can be granted subject to such terms and conditions as the Commissioner may therein determine and for a period not exceeding one year, but it can be renewed for a further period or periods not exceeding one year at any one time. Any such permit can at any time be revoked by the Commissioner or the Inspector of Mines for good cause shown. An annual fee is payable in advance in respect of every quarry permit (2).

According to Article 39 (1) ML quarry permits can be granted subject to such terms and conditions and in respect of such areas and subject to the payment of such rentals and fees as may be determined by the Governor. Quarry licences can be granted for such period not exceeding 25 years and can be renewed for a further period or periods not exceeding 25 years at any one time (2).

8.3.1.3 Authorities

The main responsible authority for mining under the responsible Minister is the Inspector of Mines (Article 30 ML).

The Inspector of Mines or other person authorized by the Governor can enter upon any land on which mining is being carried out or which is the subject of any mining lease, and inspect such mining or works in connection therewith. He can inspect any registers, books and plans connected with such mining (Article 30 ML). If at any time it is shown to the satisfaction of the Governor that a mine is in such condition as to render mining dangerous to the safety or health of persons employed in such mine, the Governor can order that such mine is closed (Article 31 ML).

8.3.1.4 Fees and Taxes

All minerals or quarry materials obtained in the course of any prospecting, mining or quarrying are liable to royalties (Article 11 ML).

8.3.2 Additional Legal Basics

Table 30: List of Acts significant for raw materials – Cyprus

Name of Law	Law No. and Date of Issuing
Integrated Pollution Prevention and Control Law	Law No. 56 (I) of 2003
Air Quality Law	Law No. 53 (I) of 2004
Forest Law	Law No. 14 of 1967, as amended by Law No. 78A of 2003
The Minimum Requirements for Safety and Health at Work (Surface and Underground Extractive Industries) Regulations	Law No. 275 of 2002
Statistics Law	2000

9 Czech Republic

9.1 General Facts

The national territory covers 78,866 km² and the inhabitants are 10.476.543 (2009 estimation). The population density is 132/km². The GDP per capita for 2009 is estimated at $ 18.557 – a GDP per capita of 82% of the European Union average.

Constitutional Structure

The form of government is a parliamentary republic. Since 2000 the Czech Republic is divided into 13 regions and the capital city of Prague. Each region has its own elected Regional Assembly and president (hejtman).

9.2 Production of Raw Materials

Table 31: Production Data – Czech Republic (World Mining Data, Weber and Zsak, 2008)

Resources		2002	2003	2004	2005	2006	Change 02/06	Change 05/06
Bentonite	(t)	174.000	199.000	201.000	186.000	220.000	26,44	18,28
Diatomite	(t)	28.000	41.000	33.000	38.000	53.000	89,29	39,47
Feldspar	(t)	401.000	421.000	488.000	472.000	487.000	21,45	3,18
Graphite	(t)	16.000	0	5.000	3.000	5.000	−68,75	66,67
Gypsum	(t)	108.000	104.000	71.000	25.000	16.000	−85,19	−36,00
Kaolin	(t)	803.000	914.100	849.640	854.040	828.960	3,23	−2,94
Steam-Coal	(t)	6.820.000	5.780.000	5.990.000	6.110.000	6.008.000	−11,91	−1,67
Coking-Coal	(t)	7.650.000	7.870.000	7.320.000	7.140.000	7.029.180	−8,12	−1,55
Lignite	(t)	49.335.000	50.390.000	48.290.000	49.125.000	48.915.000	−0,85	−0,43
Nat. Gas	(Mm³)	91	131	175	356	148	62,64	−58,43
Oil	(t)	253.000	310.000	299.000	306.000	259.000	2,37	−15,36
Uranium	(t)	562	540	513	482	422	−24,91	−12,45

Table 32: Aggregates Production – Annual Statistics/Czech Republic, 22 April 2008, quantities in million tonnes (UEPG 2008)

Sand & Gravel (1)		Crushed rock (2)		Marine Aggregates		Recycled Aggregates (3)		Manufactured Aggregates (4)	
2005	2006	2005	2006	2005	2006	2005	2006	2005	2006
25,5	27,1	38,0	41,5	n.a.	0,0	3,4	3,8	0,3	0,3

(1) Sand and Gravel: sold production including crushed gravel
(2) Crushed rock: sold production (excluding crushed gravel)
(3) Recycled Aggregates: materials coming from construction and demolition waste used in aggregates market
(4) Manufactured aggregates include blast-furnace-slag, electric-arc-furnace-slag, incinerator bottom ash (IBA), pulverised fuel ash (PFA)

9.3 Normative Basics

9.3.1 Primary Legal Basics

The primary legal basics of mineral extraction activity is the Mining Law (Mining Act) No. 44 of 1988. Besides that Law No 61 of 1988 on Mining Operations, Explosives and the State Mining Administration, as amended by Law No. 274 of 2008 is to be mentioned.

9.3.1.1 General Rules

Table 33: Structure of the Mining Law – Czech Republic

Part I Basic Provisions	Article 1 Introductory Provisions	
	Article 2 Minerals	1–2
	Article 3 Classification of minerals as reserved and non-reserved	1–3
	Article 4 Mineral deposit	
	Article 5 Mineral resources	1–2
	Article 5a Organisations	

9.3 Normative Basics

	Article 6 Reserved deposit	1–2
	Article 7 Deposit of non-reserved minerals	
Part II **Obligations of organizations in the exploitation of a reserved deposit**	**Article 8**	
	Article 9 Deleted	
	Article 10	1–2
Part III **Deposit exploration and management of reserved deposits**	**Article 11** **Prospecting for and exploration of reserved deposits**	**1–6**
	Article 12 Notification of a natural accumulation of a reserved mineral	
	Article 13 Reserves of a reserved deposit and the conditions for their exploitability	1–2
	Article 14 Classification of reserves in a reserved deposit, assessment and endorsement of calculations of reserves in reserved deposits	1–4
	Article 14a Depreciation of reserved deposit reserves	1–4
	Article 14 b Proposal for depreciation of reserved deposit reserves	1–3
	Article 14c Decision to depreciate reserves in a reserved deposit	1–6
Part IV **Protection of mineral resources**	**Article 15** **Provision for protection of mineral resources during territorial planning work**	**1–2**
	Article 16 Protected deposit area	1–3
	Article 17 Establishment of a protected deposit area	1–8

	Article 18 Limitation of some activities within a protected deposit area	1–2
	Article 19 Placing of buildings and facilities within a protected deposit area	1–2
	Article 20 Land management	1–3
Part V **Title deleted**		
Part VI **Construction of mines and quarries**	**Article 23** **Design, construction and reconstruction of mines and quarries**	**1–6**
Part VII **Extraction of reserved deposits**	**Article 24** **Authorization for extraction of a reserved deposit**	**1–11**
	Article 25 Mining claim	1–3
	Article 26 Boundaries of mining claim	1–3
	Article 27 The delimitation, alteration and abolishing of a mining claim	1–9
	Article 28 proceedings for the delimitation, alteration and abolishing of a mining claim	1–9
	Article 29 Records	1–6
	Article 30 Economical exploitation of reserved deposits	1–8
	Article 31 Duties and rights of an organization during extraction of reserved deposits	1–6
	Article 32 Plans for opening, preparation and extraction of reserved deposits and plans for safeguarding and liquidation of principal mine works and quarries	1–5
	Article 32a Fees	1–10

	Article 32b deleted	
	Article 33 Solution of conflicts of interests	1–8
Part VIII Other interventions affecting the Earth's crust	**Article 34** Special interventions affecting the Earth's crust	**1–4**
	Article 35 Old mine works	1–7
Part IX Damages caused by mining and compensation for damages	**Article 36** Damages caused by mining	**1–3**
	Article 37 Compensation for damages caused by mining	1–7
	Article 37a Creation of financial funds	1–6
Part X Common provisions	**Article 38** Safety of operations	
	Article 39 Mining survey and geological documents	1–3
	Article 40 Mine water	1–4
	Article 41 Relationship to administrative rules	
	Article 41a	
	Article 42 Deleted	
Part XI Temporary and concluding provisions	**Article 43** Temporary provisions	**1–5**
	Article 44 Abolishing provisions	1–2
	Article 45 Legal force	

	Article 44a Annulment of provisions valid from the date Czech National Council Act No. 541/1991 Coll. comes into legal force	Part I Part II
	Article 43a Temporary provisions on measures valid from the date when Czech National Council Act No. 541/1991 Coll. came into legal force	1–8

Minerals are classified as reserved[1] and non-reserved minerals according to Article 3 ML:

(1) Reserved minerals are the following:

a) radioactive minerals,
b) all kinds of coal, crude oil and flammable natural gas and bituminous rock,
c) minerals from which metals can be produced by industrial processes,
d) magnesite,
e) minerals from which phosphorus, sulphur and fluoride or their compounds can be produced by industrial processes,
f) rock salt, potassium, boron, bromine and iodine salts,
g) graphite, barite, asbestos, mica, talc, diatomite, glass-making and welding sands, mineral dyes, bentonite,
h) minerals from which rare earth elements and elements with semiconductor properties can be produced by industrial processes,
i) granite, granodiorite, diorite, gabbro, diabase, serpentine, dolomite and limestone provided they are suitable for quarrying and polishing, and travertine,
j) technically utilizable mineral crystals and precious stones,
k) halloysite, kaolin, ceramic clays, fireclays and claystones, gypsum, anhydrite, feldspar, perlite and zeolite,

---

1 Regarding the term '*reserved*': According to Article 6 ML: If a reserved mineral is found in a quantity and quality indicating a reasonable expectation of the accumulation thereof, the Ministry of the Environment of the Czech Republic has to issue a reserved deposit certificate. The reserved deposit certificate is to be sent by the Ministry of the Environment of the Czech Republic to the Ministry of the Industry and Trade of the Czech Republic and to the Regional Authority, to the District Mining Authority, to the territorial planning authority, to the construction office and to the organization on whose behalf the prospecting for and exploration of the reserved deposit was undertaken.

l) quartz, quartzite, limestone, dolomite, marl, basalt, phonolite, trachyte provided that these minerals are suitable for chemical and technological processing or smelting,
m) mineral water from which reserved minerals can be produced by industrial processes,
n) technically utilizable natural gases which are not listed under letter b).

Other minerals are non-reserved minerals (2).

Ownership of minerals rights

Mineral resources within the territory of the Czech Republic are owned by the Czech Republic (Article 5 ML).

9.3.1.2 Issuing of Permits

Prospecting and exploration

Prospecting for and exploration of reserved mineral deposits and deposits of non- reserved minerals can be carried out by the operator in exploration areas that are established pursuant to the regulations of this law (Article 11 ML). The operator is obliged to

a) examine the reserved deposit in such a way so as to assess all potentially useful minerals,
b) examine the deposit conditions so as to enable the construction of mines and quarries, their opening, preparation and extraction of the reserved deposit to be planned and implemented in accordance with the basic principles of mining technology, and to ensure efficient utilization of the deposit.

Exploitation

Exploitation of mines and quarries is subject to general regulations Article 23 (1) ML. Documents on construction must ensure the following (2):

a) economical exploitation of the reserved deposits,
b) optimum location of surface and underground equipment, buildings and mine works and use of the most suitable mining methods,
c) depositing and storage of extracted and temporarily unused minerals and waste materials (spoil tips, heaps and tailing ponds),
e) safety of operations and health and safety protection at work, safeguarding of mine works,
f) limitation of negative environmental impacts,

g) the preparation of a complex study of the area affected by mining operations, particularly of relationships to other branches of the national economy, the effect on property owners, and on legally protected general interests.

Mining claim

Authorization of an operator to extract a reserved deposit is established by delimitation of the mining claim. The operator may commence extraction in the delimited mining claim however, only after obtaining a permit from the District Mining Authority. The District Mining Authority can link together administrative procedures for the delimitation of a mining claim and proceedings for the permit of mining activities according to the special regulations (Article 24 ML).

Mining claims are delimited on the basis of results of exploration of a deposit on the basis of the extent, position of the reserved deposit, and taking into account the reserves and positional conditions so as to facilitate economical extraction of the deposit. The surface boundaries of a mining claim are to be defined by a closed geometrical figure with straight linear sides whose apices are defined by coordinates given in a valid coordinate system. Its spatial boundaries under the surface are normally specified by vertical planes passing through the surface boundaries. The boundaries of the delimited mining claim are to be designated by the territorial planning authority in the land planning documents (Article 26 ML).

Economical exploitation of reserved deposits (Article 30 ML)

Reserved deposits must be exploited economically. Economical exploitation of reserve deposits is understood to mean the extraction, processing and benefication of extracted minerals on the basis of contemporary technical and economical conditions. The principles of mining technology, health and safety at work and safety of operations must be adhered to, and unwarranted negative impacts on the working and natural environments must be avoided. During exploitation of reserved deposits it is particularly necessary to extract reserves of a reserved deposit, including secondary minerals, as completely as possible and with minimum losses and pollution.

Duties and rights of an operator (Article 31 ML)

An operator is entitled to extract a reserved deposit within the delimited mining claim (1). If, during the extraction of a reserved mineral for which a mining claim has been delimited, a deposit of another reserved mineral is found within the mining claim, the operator is obliged without delay to inform the Ministry of the Environment of the Czech Republic, Ministry of the Industry and Trade of the Czech Republic and the District Mining Authority (2). If exploration of the deposit confirms that the detected deposit can be extracted and that its extraction by another operator would not be economical, the district mining authority may require that the operator also extract the second reserve deposit.

9.3.1.3 Authorities

The main responsible authority for mining is the Minister of Environment.

The District Mining Authority is the competent authority of controlling mineral raw resource extraction. An application regarding prospecting and exploration is to be considered by the Ministry of the Industry and Trade with the approval of the Ministry of the Environment. An application regarding extraction will be handled by the Ministry of the Industry and Trade after a consultation with the Czech Mining Authority (Article 14c ML).

9.3.1.4 Fees and Taxes

According to Article 32a (1) ML an operator is obliged to pay, into an account of the relevant District Mining Authority, an annual fee for the mining claim for each ha within its surface boundaries. The Government Order is to set the size of the fee within the range 100 CSK to 1000 CSK from ha graduated taking into account a degree of a protection of environment in the affected area. The District Mining Authority is to transfer this fee to the municipality within whose territory the mining claim is located.

An operator is obliged to pay, into an account of the District Mining Authority, an annual fee for the extracted reserved minerals or reserved minerals after their treatment and refinement carried out in connection with their extraction. This fee is to be no more than 10% of the market price of the extracted minerals. The average market price in the year in which the reserve minerals were extracted is considered decisive (2). The Ministry of the Industry and Trade of the Czech Republic in agreement with the Czech Mining Authority, the Ministry of the Environment of the Czech Republic and the relevant state administration authorities can in justified cases, particularly to assist mining operations and utilization of mineral resources, reduce the amounts of fees for extracted minerals (3).

From fees obtained pursuant to par. 2, the District Mining Authority is to transfer 25 % into the state budget of the Czech Republic and 75 % into the budget of the community within whose territory the mining space is located (4). The obligation to pay fees pursuant to par. 1 and 2 commences in the calendar year following the year in which the mining claim was delimited. The obligation to pay fees pursuant to par. 1 terminates with the abolishing of the mining claim (5). If an operator does not pay fees pursuant to par. 1 and 2, it is obliged to pay a penalty for each day of delay of 0.1 % of the amount owed. If an operator does not pay the set fees within the extended period set by the District Mining Authority, the District Mining Authority can withdraw its permit to extract the reserved deposit (6).

9.3.2 Additional Legal Basics

Table 34: List of Acts significant for raw materials – Czech Republic

Name of Law	No. of Law and Year of Issuing
Law on geological works	Law No. 62 of 1988, as amended 124 of 2008
Civil Engineering Act	Law No. 50 of 1976
Law on the environment	Law No. 17 of 1992
Environmental Impact Assessment Act	Law No. 100 of 2001
Forestry Act	Law No. 289 of 1995, as amended by Law No 167 of 2008
Nature and Country Protection Act	Law No. 114 of 1992, as amended by Law No 267 of 2006
Air Protection Act	Law No. 86 of 2002
Water Act	Law No. 254 of 2001, as amended by Law No 20 of 2004
Civil Code	Law No. 40 of 1964 as amended
Decree Coll., which lays down the amount of the hallmarking fees and the method of payment thereof as amended by Decree No. 364/2003	No. 53/1993 as amended by Decree No. 364/2003
Czech National Council Act, on Mining Operations, Explosives and State Mining Administration	No. 61/1988 of the Collection

Name of Law	No. of Law and Year of Issuing
Act on Land Development Planning and Civil Engineering Order (Civil Engineering Act)	No. 50 of 1976
Act of Legal Offence	No. 200 of 1995
Labour Code	No. 65 of 1965
Act on Waste	No. 185/2001
Regulation on the Waste Management Plan of the Czech republic.	2003
Decree of the ČBÚ on the Conditions of Non-reserved Mineral Deposit Exploitation	No. 175 of 1992, as amended by the Decree No. 298 of 2005
Decree of the MH ČR the Payment of Royalties on Mining Leases and Extracted Minerals	No. 617 of 1992, as amended by Decree of the MPO No. 63 of 2005

Water Act

The purpose of this Act is to protect surface water and groundwater, stipulate conditions for economic utilisation of water resources whilst preserving and improving the quality of surface water and groundwater. The Act regulates legal relationships involving surface water and groundwater, the relationships of natural persons and legal entities with surface water and groundwater utilisation, as well as the relationships with plots of land and buildings directly connected with these waters, in the interests of ensuring sustainable water utilisation (Article 1). According to Article 14 permit for extraction of sand, gravel from the plots on which a watercourse is located is required.

10 Denmark

10.1 General Facts

The national territory covers 43 098.31 km² and the inhabitants are 5.519.441 (2009 estimation). The population density is 127,9/km². The GDP per capita for 2009 is estimated at $ 56.115.

Constitutional Structure

The form of government is a constitutional monarchy with a parliamentary system of government. Since January 2007 (during the Danish Municipal Reform) it is divided into 5 regions and 98 municipalities and replaced the former 13 counties. Greenland and the Faroe Islands belong also to the Kingdom of Denmark, but have autonomous status and are largely self-governed.

10.2 Production of Raw Materials

Table 35: Production Data – Denmark (World Mining Data, Weber and Zsak, 2008)

Resources		2002	2003	2004	2005	2006	Change 02/06	Change 05/06
Salt	(t)	570.000	560.000	610.000	610.000	620.000	8,77	1,64
Nat. Gas	(Mm³)	8.300	8.000	8.400	9.400	10.878	31,06	15,72
Oil	(t)	17.700.000	18.000.000	19.200.000	18.719.900	17.847.496	0,83	−4,66

Table 36: Aggregates Production – Annual Statistics/Denmark, 22 April 2008, quantities in million tonnes (UEPG 2008)

Sand & Gravel (1)		Crushed rock (2)		Marine Aggregates		Recycled Aggregates (3)		Manufactured Aggregates (4)	
2005	2006	2005	2006	2005	2006	2005	2006	2005	2006
58,0	n.a.	0,3	n.a.	n.a.	13,6	n.a	n.a.	n.a.	n.a.

(1) Sand and Gravel: sold production including crushed gravel
(2) Crushed rock: sold production (excluding crushed gravel)

(3) Recycled Aggregates: materials coming from construction and demolition waste used in aggregates market
(4) Manufactured aggregates include blast-furnace-slag, electric-arc-furnace-slag, incinerator bottom ash (IBA), pulverised fuel ash (PFA)

10.3 Normative Basics

10.3.1 Primary Legal Basics

The primary legal basics of mineral extraction activity is the Mining Law ('Raw Materials Act') No. 569 of 1997, as amended by Law No. 145 of 2002.

The Danish legislation differs between onshore mineral resources and marine mineral resources, but both categories are regulated either by the Raw Materials Act or the Subsurface Act. The Raw Materials Act differs between commercial and non-commercial extraction, but both categories are regulated by the Act. The Raw Materials Act does not apply in the Faroe Islands and Greenland.

Deep seated mineral resources onshore as well as on the continental shelf are regulated by the Danish Subsurface Law No. 552 of 1995. The mineral resources falling under the Subsurface Act belong to the State.

10.3.1.1 General Rules

Table 37: Structure of the Mining Law – Denmark

Chapter 1 Purpose
Articles 1–3
Chapter 2 Management of Raw Materials
Article 4
Chapter 3 Deposits on Land
Mapping Articles 5–6
Permits, etc. Articles 7–11
Notification and Appeal Articles 12–17
Chapter 4 Deposits in Territorial Waters and on the Continental Shelf
Mapping and Planning Article 18
Authorization of Extraction Equipment Article 19
Permits and Demarcation of Extraction Areas Articles 20–22
EIA Procedure Article 23

Lapse, Revocation or Amendment of Permits Article 24
Installations on the Continental Shelf Article 25
Appeals Article 26
Chapter 5 Expropriation
Article 27
Chapter 6 Reports and Surveys
Articles 28–29
Access to Carry out Surveys on Public or Private Property Article 30
Chapter 7 Supervision
Articles 31–33
Chapter 8 Administration
Articles 34–36
Reporting Obligation Article 37
Delegation of the Powers of the Minister Article 38
Expert Committees Article 39
Fees Article 40
International Obligations Article 41
The Municipalities of Copenhagen and Frederiksberg Article 42
Chapter 9 Legal Proceedings and Penalties
Articles 43–45
Chapter 10 Entry into Force and Transition Provisions, etc
Articles 46–53

The Act covers stone, gravel, sand, clay, lime, chalk, peat, and similar deposits (Article 2 ML).

Ownership of mineral rights

The land-based minerals belong to the landowner, marine resources on the sea floor are owned by the State.

Acquiring mineral rights

Exploitation rights of marine resources are acquired through a permit system under the Raw Materials Act. A permit will allow the operator to extract minerals from a designated area. The extraction of land-based minerals requires a private contract between the landowner and the operator. The landowner must pay full compensation.

10.3.1.2 Issuing of Permits

Exploration

The Ministry of the Environment or the County councils can initiate prospecting and exploration regardless of the land owner. The land owner and/or land users must be informed about the planned activities at least two weeks before the work starts. The land owner is entitled to full compensation for any damage caused by the exploration activities. The Ministry (Geological Survey of Denmark and Greenland) is responsible for exploration work on marine mineral resources. Private parties can initiate exploration work on private land, provided they have an agreement with the landowner. However, a permit is needed from the County council if the investigation encompasses trial extraction or sampling of more than 200 m^3 of material. Application for an exploration permit is made using a standard form issued by the Ministry of the Environment.[1]

The results and findings (including samples) made in connection with any type of private exploration activity must be reported to the County council within three months after the work has been completed. The Council passes the information on to the Ministry of the Environment and GEUS.

Extraction

Extraction of raw materials and establishment of installations at the extraction site for such extraction activity is subject to a permit from the county council. Permits for extraction on beaches and other coastal stretches with no continuous land vegetation are granted only with the consent of the Danish Coast Authority (Article 7 ML).

An application for a permit pursuant to Article 7 shall also be regarded as an application for a permit pursuant to other legislation concerning extraction of raw materials from the soil (Article 8 ML; see also below, consultation process).

Permits for extraction of raw materials can be granted for a period of up to ten years. In special cases the permit may be granted for a longer period. Other raw materials than stated in the permit may be extracted in the area covered by the permit, if reported to the county council within four weeks of commencement of such extraction. Within four weeks of receipt of the report the county council may prohibit any further extraction which infringes the permit granted, or specify particular conditions for extraction (Article 9 ML).

---

1 Nielsen, K. (2004), Country report Denmark, in: Department of Mining and Tunnelling, University of Leoben, Minerals Policies and Supply Practices in Europe.

According to Article 10 ML a permit pursuant to Article 7 (1) contain conditions for:

- the operation of the enterprise and for after-treatment of the area in order to limit environmental damage and prevent pollution of groundwater and soil;
- the provision of security for the after-treatment of the extraction area;
- extraction and after-treatment to take place in accordance with a plan approved by the county council and which contains the main elements of the extraction and after-treatment; and
- the processing of the raw materials in the best possible way in relation to their quality or their use for specific purposes.

The security provided as stated in Article 7 (1) must cover the county council's costs of after-treatment by execution of action to which a person is obliged, on that person's account. The county council may adjust the size of the security bond should the basis for calculation change significantly Article 10 (3). The county council must undertake that the after-treatment conditions are registered in the land registry in respect of the property concerned at the owner's expense Article 10 (5). A permit shall lapse if it is not utilized within three years after it is granted or has not been used for three consecutive years Article 10 (6).

Consultation Process

Commercial mineral extraction and erection of buildings and plants needed for the exploitation of a deposit can only commence after the County council has issued an extraction permit. Permits are normally granted for ten years, but can be extended by applying for a new permit. Application for an extraction permit is made using a standard form issued by the Ministry of the Environment. Permits are subject to public hearings before activities can commence. The hearing period and term of objection shall be at least eight weeks. One very important aspect with regard to the permitting process for extraction is the fact that Article 8 ML specifically states that an application for mineral extraction is also an application for permits regarding other relevant legislation e. g. the Water Supply Act, the Nature Conservation Act etc. This means that the county authorities will assess and handle all the relevant legal issues in co-operation with other responsible authorities as part of the permitting process.[2]

2　Ibidem.

Restoration

When the county grants a permit to develop an extraction site, a restoration plan must be prepared. The plan may be submitted in connection with the application for a permit, or it can be developed afterwards. The plan can either be prepared by the county administration at the cost of the operator, or by the operator for approval by the county. The contents and objectives of a restoration plan will depend on the type of operation and the planned land use after extraction has been completed. A financial guarantee must be given to ensure that restoration will be made according to the plan accepted by the county authority.[3]

10.3.1.3 Authorities

The main responsible authority for mining is the Minister for Environment and Energy (Forest and Nature Agency).

According to Article 31 ML the county council has to ensure extraction of raw materials on land, and that conditions attached to permits are met (1). As concerns extraction and exploration of raw materials in territorial waters and on the continental shelf, the Minister for Environment and Energy shall ensure compliance with the Act is met (2). The Minister for Environment and Energy can make regulations concerning the execution of the county council's supervision (5). The Minister for Environment and Energy can determine that other authorities shall carry out the supervision (6).

The Minister for Environment and Energy and the county council or persons authorized by these authorities have the right of access without court order to public and private properties and sites, or other extraction equipment in order to exercise the powers provided by this Act (Article 32).

The Forest and Nature Agency under the Ministry of the Environment is responsible for mineral exploitation onshore.

Monitoring and inspection of the commercial mineral operations is done by the county authorities on a regular basis, and the Raw Materials Act has very explicit formulations about the right to make inspections, gain access etc. Inspection related to worker safety and health conditions is made by the Work Inspection Authority which is an agency under the Ministry of Employment.

---

[3] Ibidem.

10.3.1.4 Fees and Taxes

According to Article 40 ML the Minister for Environment and Energy may make regulations on fees to cover all or part of the expenses of the authorities in the administration and supervision.

A general fee is paid to the State on each volume unit of mineral extracted, and the fee shall cover the administrative costs of the Forest and Nature Agency in connection with the Raw Materials Act. The fee is partially also seen as one of various policy tools that have been introduced in order to encourage recycling of heavy building waste materials.

There is no extra fee or royalty paid to the State for extraction of marine resources from the sea floor.

Compensation to the land owner for the value of extracted mineral is a matter of negotiation between the parties. The landowner shall also have full compensation for any damage caused by exploration, extraction and/or processing.

10.3.2 Additional Legal Basics

Table 38: List of Acts significant for raw materials – Denmark

Name of Law	No. of Law and Year of Issuing
Spatial Planning Act	Law No. 763 of 2002
Nature Protection Act	Law No. 835 of 1997
Environment Protection Act	Law No. 698 of 1998
Subsurface Act	Law No. 552 of 1995
Forest Act	Law No. 383 of 1989, as amended by Law No. 453 of 2004
Building Act	Law No. 452 of 1998
Water Supply Act	Law No. 130 of 1999
Act on Statistics Denmark	No. 599 of 2000
The Planning Act in Denmark Consolidated Act	No. 763 of 2001
Consolidated Competition Act	No. 785 of 2005
Act on Access to Information on the Environment	No. 292 of 1994
Act on Waste Deposits	No. 420 of 1990
Act (Exclusive Economic Zones Act)	No. 394 of 1996
Contaminated Soil Act	No. 370 of 1999

Environment Protection Act

The purpose of this Act is to contribute to safeguarding nature and environment, thus enabling a sustainable social development in respect for human conditions of life and for the conservation of flora and fauna (Article 1). In the administration of this Act weight is given to the results achievable by using the least polluting technology, including least polluting raw materials, processes and plants and the best practicable pollution control measures (Article 3). In the design and operation of plants, including raw materials measures must be taken to minimize the use of resources, pollution and generation of waste (Article 4).

Water Supply Act

The purpose of this Act is to ensure that the exploitation and the thereto associated protection of water reserves are carried out based on comprehensive planning (Article 1). The administration of this Act shall emphasize the quantity of water reserves, the needs of the population and businesses for water supply that is adequate and satisfactory in quality, protection of the environment and of nature, including the use of raw material reserves (Article 2).

Forest Act

The purpose of the Act is to conserve and protect Danish forests (Article 1).

11 Estonia

11.1 General Facts

The national territory covers 45,228 km² and the inhabitants are 1.340.415 (2009 estimation). The population density is 29/km². The GDP per capita for 2009 is estimated at $ 14.266.

Constitutional Structure

The form of government is a democratic parliamentary republic. It is divided into 15 counties (Maakonnad) which are further divided into municipalities of two types: urban or rural municipalities. Each municipality is a unit of self-government

11.2 Production of Raw Materials

Table 39: Production Data – Estonia (World Mining Data, Weber and Zsak, 2008)

Resources		2002	2003	2004	2005	2006	Change 02/06	Change 05/06
Oil-Shale	(t)	10.513.000	12.459.000	11.328.000	11.310.000	11.977.100	13,93	5,90

Table 40: Production of aggregates – Estonia. (Metric tons unless otherwise specified), (USGS, 2008)

		2002	2003	2004	2005	2006
Clays:						
For brick	1000 m³	149.200	134.900	136.600	151.800	n.a.
For cement	1000 m³	19.000	27.300	31.600	37.200	n.a.
Dolomite:						
For building	m³	n.a.	291.200	323.400	261.700	n.a.
For finishing	m³	n.a.	3.200	1.300	2.000	n.a.

		2002	2003	2004	2005	2006
For industry (technological limestone)	m³	n.a.	150.800	171.900	155.300	n.a.
Gravel, pebbles, shingle and flint	m³	n.a.	n.a.	n.a.	597.100	n.a.
Lime Limestone:						
For building	m³	n.a.	1.255.000	1.547.000	1.922.000	n.a.
For cement	m³	366.200	372.200	430.500	335.100	n.a.
For industry (technological limestone)	m³	n.a.	62.500	93.900	86.300	n.a.
Sand and gravel	1000 m³	2.033	4.470	3.131	2.186	n.a.
Silica sand, industrial		22.500	41.300	49.800	53.800	38.000

11.3 Normative Basics

11.3.1 Primary Legal Basics

The primary legal basics of mineral extraction activity is the Mining Law No. 20, 118 of 2003 amended by the Law No. 18, 131 of 2004.

Besides that there exist the Earth's Crust law No 84, 572 of 2003 as amended by Law No. 15, 87 of 2005. The Earth's Crust law provides for the procedure for and the principles of exploration, protection and use of the earth's crust, with the purpose of ensuring economically efficient and environmentally sound use of the earth's crust. This Act regulates: 1) geological investigation; 2) geological exploration; 3) extraction of mineral resources, except in the part regulated by the Mining Act; 4) the rights of the owner of an immovable upon use of mineral resources within the boundaries of the owner's immovable; 5) restoration of the land disturbed by geological investigation, geological explorations or mining.

11.3.1.1 General Rules

Table 41: Structure of the Mining Law 2003, Estonia

Chapter 1 General Provisions	Article 1. Scope of application of Act	Item 1–3
	Article 2. Application of other Acts	Item 1–4
	Article 3. Definitions for purposes of this Act	Item 1–5
Chapter 2 Mining and Secondary Utilisation of Workings	Article 4. Safety requirements	Item 1–5
	Article 5. Suspension and termination of mining	Item 1–4
	Article 6. Permit for secondary utilisation of underground workings, and building permits and permits for use of underground structures	Item 1–2
	Article 7. Plans for mining and secondary utilisation of workings	Item 1–3
	Article 8. Mine survey operations and mine surveying documentation	Item 1–5
Chapter 3 Holders of Exploration and Extraction Permits, and Undertakings	Article 9. Holders of exploration and extraction permits	Item 1–2
	Article 10. Right to operate as undertaking	Item 1–2
	Article 11. Obligations of undertakings engaged in mining or secondary utilisation of workings	Item 1–7
	Article 12. Obligations of undertakings engaged in preparation of plans	Item 1–5
	Article 13. Mandatory documentation for undertakings engaged in mining or secondary utilisation of workings	Item 1–5
Chapter 4 Specialist in Charge	Article 14. Specialist in charge	Item 1–2
	Article 15. Requirements for specialists in charge in certain areas of activity	Item 1–4
	Article 16. Duties of specialist in charge	Item 1–2
Chapter 5 Authority Assessing and Attesting Conformity of Persons	Article 17. Assessment and attestation of conformity of persons	Item 1–3
	Article 18. Authority assessing and attesting conformity of persons	Item 1–4
	Article 19. Liability insurance	Item 1–2

Chapter 6 **Registration of** **Undertakings**	Article 20. Registration application	Item 1–3
	Article 21. Registration procedure and registry data	Item 1–2
	Article 22. Refusal to register	
	Article 23. Deletion of registration	
	Article 24. (Repealed – 10. 08. 2004 entered into force 15. 04. 2004 – RT I 2004, 18, 131)	
Chapter 7 **State Supervision**	Article 25. State supervisory authority	
	Article 26. Competence of Inspectorate	Item 1–2
	Article 27. Rights and obligations of officials exercising state supervision	Item 1–4
	Article 28. Precept and decision	Item 1–4
	Article 29. Contestation of precept or act	Item 1–4
	Article 30. Inspection of plans	Item 1–3
Chapter 8 **Liability**	Article 31. (Repealed – 10. 08. 2004 entered into force 15. 04. 2004 – RT I 2004, 18, 131)	
	Article 32. Violation of requirements for mining, secondary utilisation of underground workings and preparation of related plans	Item 1–2
	Article 33. Violation of requirements for assessment and attestation of conformity of persons	Item 1–2
	Article 34. Proceedings	Item 1–3
Chapter 9 **Implementing** **Provisions**	Article 35. Transitional provisions	Item 1–4
	Article 36. Amendment of Rescue Act	
	Article 37. Amendment of Earth's Crust Act	Item 1–2
	Article 38. Amendment of State Fees Act	Item 1–3
	Article 39. Repeal of Technical Supervision Act	
	Article 40. Entry into force of Act	Item 1–3

Minerals classification

The mineral deposits are divided (according to their value) into the deposits of state importance and local importance. The list of the deposits of state importance is certified by the Government of the Republic on behalf of the Minister of the Environment.[1]

Ownership of minerals

According to Article 4 of the Earth's Crust Law the ownership of mineral resources relating to bedrock minerals, mineral resources in mineral deposits of national importance and lake mud and sea mud (medicinal mud) belong to the State and the immovable property ownership of other persons does not extend to these. Selected minerals belong to the State, for example bedrock clay, dolomite, phosphorite, limestone and sand used in technological processes (technological sand). Other minerals belong to the land owner.[2]

Acquiring mineral rights

The mineral rights are granted exclusively by the permit for exploitation.

11.3.1.2 Issuing of Permits

Exploration

Procedure for exploration – Earth's Crust Law (Article 8)

The procedure for exploration is established by the Minister of the Environment. The following must be established:

1) the requirements for geological mapping;
2) the requirements for exploration;
3) the extent of activities which change the condition of soil and the earth's crust;
4) the requirements for geological mapping and for the preparation of an exploration report.

Exploitation

According to Article 9 ML holders of an (exploration or) extraction permit must ensure that the plans for mining are prepared and the mining operations are conducted with the conditions set by the (exploration or) extraction permit.

An operator must ensure compliance with the requirements arising from this Act, by performing the mining operations in compliance with the plans prepared according to the relevant requirements (Article 11).

---

1 Koziol, W., Kawalec, P. (2004), Country Report Estonia, in: Department of Mining and Tunneling, lc.
2 Ibidem.

An operator engaged in the preparation of plans has to ensure that the plans conform to the primary purpose; to ensure that the plans are prepared by persons with sufficient professional training therefore; to preserve in full all plans prepared thereby for at least 7 years as of the termination of work carried out on the basis of such plans (Article 12).

Mandatory documentation for operators engaged in mining (Article 13)

An operator must also provide the following documents: 1) a plan for mining; 2) documents concerning risk assessment; 3) occupational safety instructions and instructions for the use of equipment and machinery; 4) mine surveying documentation (1).

In addition to the documents specified in subsection (1) of this section, an operator engaged in underground work must also provide the following documents provide: 1) a plan to remedy the effects of accidents; 2) a procedure for keeping records regarding persons staying underground; 3) dewatering, ventilation and electrical supply plans, and other technological plans (2).

An operator has to prepare a development plan for each calendar year or for a longer period divided into years and to submit the plan to the local government of the location of the mining site for information purposes if the local government so required. The purpose of a development plan covering the activities of an operator is to ensure effective control over the environmental impact of the mining operations and to ensure the optimum scope of the operations (3).

The requirements for the mandatory documentation for operators are established by the Minister of Economic Affairs and Communications (4).

Registration of operators

According to Article 20 ML an operator must submit a registration application to the authorised processor of the state register of operators operating in areas of activity subject to special requirements. A registration application must set out the following: 1) the name, address and other contact details of the operators, and the registry code or personal identification code; 2) the area of activity (assessment and attestation of the conformity of persons, mining, or preparation of plans) in which the person wishes to operate.

11.3.1.3 Authorities

The main responsible authority for mining is the Minister of Environment. State supervision over conformity with the requirements provided in the mining law will be exercised by the Technical Inspectorate (Article 25 ML).

The following are within the competence of the Inspectorate (Article 26 (1) ML): 1) supervision over the safe organisation of work at sites and over the operation of authorities assessing and attesting the conformity of persons; 2) investigation of the causes of breakdowns or accidents on sites; 3) issue of precepts and making of resolutions; 4) inspection of plans. The Director General of the Inspectorate can form committees of experts to resolve issues related to supervision. A committee of experts has an advisory role (2).

According to Article 30 (1) ML the Inspectorate has the right to inspect the compliance of plans prepared for mining operations or secondary utilisation of underground workings with the requirements. For the purpose of the inspection provided for in subsection (1), the Inspectorate has the right to involve competent persons (experts) who shall present their written opinion, or, in justified cases, to order the assessment services which are necessary for inspection (2).

11.3.1.4 Fees and Taxes

The extraction tax (royalty), stated by the Government is a tax for the right to exploit the state minerals. The extraction tax depends on the location site, nature protectional meaning and quality of the minerals. The Government of the Republic has the right to increase, decrease the rate of extraction tax and even make an exploitation free of charge.

The extraction tax of the deposits of state importance is paid in extent of 30% into the Environmental Fund and in extent of 70% into the local budget. The extraction tax of the deposits located in transboundary watercourses, territorial and coastal waters and exclusive economic zone is paid into the Environmental Fund. The extraction tax of the deposits of local importance is paid into the local budget. The local government can offer tax allowances to the payment into the local budget and free the licensee from the tax.[3]

---

3 Koziol, W., Kawalec, P. (2004), Country report Estonia, in: Department of Mining and Tunnelling, University of Leoben, Minerals Policies and Supply Practices in Europe.

11.3.2 Additional Legal Basics

Table 42: List of Acts significant for raw materials – Estonia

Name of the Law	No. of Law and Year of Issuing
Environmental Impact Assessment and Environmental Management System Act	Law No. 15, 87 of 2005
The Planning and Building Act	Law No. of 1995
Act on Sustainable Development	Law No. of 1995
Nature Conservation Act	Law No. 38, 258 of 2004 as amended by Law No. 3,15 2008
Administrative Procedures Act	Law No. of 2002
Waste Management	Law No. 57, 861 of 1998, as amended by Law No. 88, 594 of 2003
Technical Supervision Act	Law 1998
Civil Code	
Land Register Act	consolidated text 2007
Law of Property Act Implementation Act	
Competition Act	2001 (consolidated 2006)
Land Tax Act	
Water Act	
Occupational Health and Safety Act	1996
The Planning and Building Act	1995
Minister of Environment Regulation Maximum limits of hazardous substances in soil and ground water	No. 58/1999

Environmental Impact Assessment and Environmental Management System Act

This Act provides legal bases and procedure for assessment of likely environmental impact, organisation of eco-management and audit scheme and legal bases for awarding eco-label in order to prevent environmental damage (Article 1). Activities with significant environmental impact are open-cast mining where the surface of the site exceeds 25 hectares, or peat extraction where the surface of the site exceeds 150 hectares, or underground mining (Article 6 (1) 28). The operator is required to analyse, whether the activities of exploration and mining or termination of mineral resources have significant environmental impacts.

Nature Conservation Act

The purpose of this Act is to protect the natural environment by promoting the preservation of biodiversity through ensuring the natural habitats and the populations of species of wild fauna, flora and fungi at a favourable conservation status; to promote the sustainable use of natural resources (Article 1). The Nature Conservation Act refers to the Earth's Crust Law. According to Article 90 of the Earth's Crust Law a permit for extraction of mineral resources concerning lands where a protected natural object is situated must be issued on the bases provided by the Nature Conservation Act.

12 Finland

12.1 General Facts

The national territory of Finland covers 338.145 km² and the inhabitants are 5.3 million. The population density is 16/km². The GDP per capita amounts to $ 44.491 in 2009.

Constitutional Structure

The form of government is a republic. It is divided into the national level, provincial level (11 provinces) and 380 municipal governments of which 84 are cities.

12.2 Production of Raw Materials

Table 43: Production Data – Finland (World Mining Data, Weber and Zsak, 2008)

Resources		2002	2003	2004	2005	2006	Change 02/06	Change 05/06
Chromium	(t)	283.000	274.500	275.000	285.500	290.000	2,47	1,58
Nickel	(t)	3.500	3.229	3.400	3.400	2.800	−20,00	−17,65
Cadmium	(t)	4	0	0	0	0	−100,00	
Copper	(t)	14.400	14.900	15.500	15.000	13.000	−9,72	−13,33
Mercury	(t)	51	25	24	34	23	−54,90	−32,35
Zinc	(t)	34.100	39.850	37.200	40.803	35.700	4,69	−12,51
Gold	(kg)	1.600	1.550	1.360	1.296	1.300	−18,75	0,31
Platinum	(kg)	508	461	705	800	800	57,48	0,00
Silver	(kg)	29.900	33.960	49.400	47.500	50.800	69,90	6,95
Feldspar	(t)	40.000	59.362	57.149	58.000	60.000	50,00	3,45
Phosphate	(t)	800.000	810.000	850.000	823.000	800.000	0,00	−2,79
Sulphur	(t)	288.879	341.087	360.557	350.000	402.000	39,16	14,86
Talc	(t)	477.000	501.658	482.000	542.000	550.000	15,30	1,48

Table 44: Aggregates Production – Annual Statistics/Finland, 22 April 2008, quantities in million tonnes (UEPG, 2008)

Sand & Gravel (1)		Crushed rock (2)		Marine Aggregates		Recycled Aggregates (3)		Manufactured Aggregates (4)	
2005	2006	2005	2006	2005	2006	2005	2006	2005	2006
53,0	54,0	45,0	46,0	n.a.	0,0	0,5	0,5	n.a.	0,0

(1) Sand and Gravel: sold production including crushed gravel
(2) Crushed rock: sold production (excluding crushed gravel)
(3) Recycled Aggregates: materials coming from construction and demolition waste used in aggregates market
(4) Manufactured aggregates include blast-furnace-slag, electric-arc-furnace-slag, incinerator bottom ash (IBA), pulverised fuel ash (PFA)

12.3 Normative Basics

12.3.1 Primary Legal Basics

The primary legal basics of mineral extraction activity related to metal ores and industry minerals is the Mining Law No. 503 of 1965 as amended in 1997.

Construction minerals are regulated by the Land Extraction Act No. 555 of 1981. The Land Extraction Act governs extraction permits of non-claimable minerals (rock, gravel, sand, clay and soil) if the area in question is not covered by municipal development plans. The purpose of the Act is to ensure that the extraction activities that fall under the act are conducted in a way that supports protection and sustainable development of the environment.

12.3.1.1 General Rules

Table 45: Structure of the Mining Act of Finland

Chapter 1: General Provisions Articles 1–4
Chapter 2: Claim Articles 5–11
Chapter 3: Claim Right Articles 12 Article 20
Chapter 4: Appropriation of the Mining District Articles 21–39
Chapter 5: Mining Right Articles 40–51
Chapter 6: Mining Register Articles 52–55
Chapter 7: Safety Regulations Articles 56–58

Chapter 8: Surveillance of the Observance of the Act Articles 59–65

Chapter 9: Application for Alteration Articles 66–69)

Chapter 10: Miscellaneous Provisions Articles 70–79

Extractable minerals (Article 2 ML):

a) lithium, rubidium, caesium, beryllium, magnesium, strontium, radium, boron, aluminium, scandium, yttrium, the rare earth metals (lanthanides), actinium, thorium, uranium, and other actinides, germanium, tin, lead, arsenic, antimony, bismuth, sulphur, selenium, tellurium, copper, silver, gold, zinc, cadmium, mercury, gallium, indium, thallium, titanium, zirconium, hafnium, vanadium, niobium, tantalum, chromium, molybdenum, tungsten, manganese, rhenium, iron, cobalt, nickel and also platinum and the other metals of the platinum group;

b) graphite, diamond, corundum, quartz, bauxite, olivine, cyanite, andalusite, sillimanite, granate, wollastonite, asbestos, talc, pyrophyllite, muscovite, vermiculite, kaolin, feldspar, nepheline, leucite, scapolite, apatite, baryte, calcite, dolomite, magnesite, fluorspar and cryolite;

c) precious stones; and

d) marble and soapstone.

Of the extractable minerals, iron, aluminium, quartz and feldspar may be sought, claimed and exploited only if they occur in bedrock.

Ownership of minerals

The state has the responsibility about issuing mineral rights in terms of exploration and extraction of metal ores and industrial minerals. The right according to Article 1 ML to seek, claim and exploit on one's own or another's land a deposit that contains extractable minerals referred to in Article 2 ML belongs to:

(1) any natural person domiciled within the European Economic Area;
(2) any Finnish corporation or foundation; and
(3) any foreign corporation or foundation which has been established in accordance with the laws of a state belonging to the European Economic Area and which has its registered office, central administration or principal place of business in a state belonging to the European Economic Area.

The Ministry of Trade and Industry grants a permit to conduct operations governed by this act.

According to Article 3 ML everyone has the right to make geological and geophysical observations and other measurements that are to be considered necessary in seeking extractable minerals (prospecting).

A person eligible to claim has the right to reserve for himself priority to claim any mineral deposit within a stated area of up to 9 square kilometres (reservation) by making a notice of reservation to the register office of the locality where the deposit lies. The report must include the name, occupation and address, and the geographical location (applicant) of the reservation has to be specified. The location and boundaries of the district shall be marked out on the map appended to the notice (Article 7 (1) ML).

Reservation shall not apply to any area that lies nearer than 1 kilometre from an area for which another person has applied for claim, or from a claim or concession belonging to another person (Article 7 (2) ML).

The register office must within 7 days of the handing of the notice to forward the same to the Ministry of Trade and Industry.

The claim can comprise an area measuring up to 1 km². The claim must be a coherent area (Article 5 (1) ML).

If the claim application meets the conditions enacted in this act, the Ministry of Trade and Industry may give the applicant a prospecting permit for the area referred to in the application or for that part of the area for which no claim impediments exist (Article 10 (1) ML). With regard to the estimated volume and nature of requisite investigations, the claim is valid for at least 2 but not more than 5 years from the day of issue of the prospecting permit (Article 10 (2) ML).

12.3.1.2 Issuing of Permits

Prospection and exploration

The claimant has the right to carry out exploration work on his claim to ascertain the nature and extent of the deposit and, according as needed, to use ground outside the claim for roads, power lines and water or other pipelines (Article 12 (1) ML).

The claimant has the right to transfer his claim right to another person eligible to claim. The approval thereof must be recorded in the transfer and the recipient prospecting permit (Article 13).

According to Article 19 (1) ML the claimant must within a year from the relinquishment or loss of the claim right submit to the Ministry of Trade and

Industry a detailed report on the exploration work carried out on the claim area and the result thereof, with the pertinent maps appended thereto.

In addition, a representative portion of the drill cores must be sent within 5 years of the expiry of the claim to the Geological Survey of Finland for storage in the national drill core depot.

If despite systematic exploration during the period mentioned in Article 21 (1) ML, sufficient clarity has not been achieved on the possibilities of exploiting the deposit, the Ministry of Trade and Industry can, upon application made before the stipulated time has run out, grant an extension of up to 3 years (Article 21 (2) ML).

Exploitation phase

Application for a mining concession must be made in writing to the Ministry of Trade and Industry (concession application). If the application is not made simultaneously with the claim application, it must nevertheless be made during the validity of the prospecting permit, at the risk of the claim right otherwise being forfeit (Article 21 (1) ML).

Aricle 22 (1) ML describes the term *'mining district'*. A mining district must consist of a coherent area that meet the practical demands as to its size and form, and include part of the claim district. The mining district include areas that are necessary for the exploitation of the deposit, such as land for industry, storage, dumping and dwelling, and also land for roads, transport equipment, power lines, water pipelines and sewers as well as land in the neighbourhood of the deposit where there is earth suitable for filling-in during mining or other material required as supplementary material in processing the mine products.

Article 23a ML regulates the concession application concerning a project governed by the Law (468/94) on procedures to be applied in assessing the environmental impact (EIA-Law according Directive 97/11 EC). That means: The concession application must be accompanied by an environmental impact assessment conforming to said law.[1]

The Ministry of Trade and Industry must issue a concession certificate to the concession holder as proof of the concession right and of its entry in the mining register (Article 38 ML).

---

[1] Insofar as the assessment contains the information relating to environmental impacts needed for the application of this act, the same statement need not be supplied again. It shall appear from the decision on the application how the assessment made under the said act has been taken into consideration.

When the mining certificate has been issued, the concession holder has the right to process and utilize all the extractable minerals within the concession (mining operations) (Article 40 (1) ML). What is provided about claim right in Article 13 applies to transfer of concession rights (Article 42 ML). The transfer must be recorded in the original of the mining certificate.

Mining register

The Ministry of Trade and Industry keeps a mining register of claims and concessions. Everyone has the right for a stipulated fee to obtain extracts from the register (Article 52 ML). According to Article 53 ML the following must be entered in the mining register:

1. the name, occupation, domicile and address as well as citizenship;
2. the number of the prospecting permit, the day of its issue and its period of validity;
3. the name of the claim district extent and location as well as the extractable minerals that have been reported in the deposit;
4. the amount of the claim fee, and payments made;
5. the concession application for the concession district and the measures arising there from;
6. the granting of concession;
7. the transfer of the claim right and the concession right and also other acquisition and the grounds therefore insofar as reported in accordance with Article 54 for entry in the register;
8. the pledging as collateral of the claim right or concession right;
9. the cessation of the rights entered in the mining register;
10. the period of validity of the claim right and the concession right.

The Ministry of Trade and Industry must make public notice in the municipality concerned of entry made in the mining register that concerns (Article 55 ML):

(1) issue of a prospecting permit;
(2) directive on execution of concession;
(3) transfer of a claim right or a concession right to another; or
(4) relinquishment or loss of a claim right or a concession right or declaration of forfeiture.

Closing phase

After mining has ended the operator must ensure that the area including mine openings etc. is left in a state that ensures public safety to a satisfactory degree. All buildings and other surface installations, as well as ore brought to the surface but not used, must be removed within two years. If not, it will become the property of the landowner. The Mining Act does not regulate restoration and aftercare other than the issues discussed above. Restoration and aftercare are regulated by other legislation, and the environmental permit will lay down the requirements regarding the restoration and landscaping of the site.[2]

12.3.1.3 Authorities

The main responsible authority for mining is the Minister of Trade and Industry (Directorate of Mines).

It is the duty of the Ministry of Trade and Industry to see that the Mining Law and the directives and provisions based thereon are observed and to make inspections to that end. The Safety Technology Authority must carry on surveillance bearing on the safety of mines in accordance with this act and carry out inspections to that end (Article 59 (1) ML).

12.3.1.4 Fees and Taxes

In order to acquire the claim right, the claimant must pay the owner of ground belonging to the claim district compensation for the calendar year (Article 15 Abs. 2 ML). The claimant must within one year from the issue of the prospecting permit deliver to the Ministry of Trade and Industry proof of the payment of the first claim compensation, at the risk of the claim otherwise being declared forfeit.

For the following years the claimant must pay the claim compensation by the 15th March each year at the risk that the Ministry of Trade and Industry can otherwise, upon request of the land owner and after the claimant has been heard, declare the claim forfeit. The claimant must not carry out exploration work referred to in Article 12 within the claim before the claim compensation for the year in question has been paid.

The claimant must pay the State a claim fee for every calendar year (Article 16 (1) ML). The amount of the claim fee per surface unit is determined in decree. It shall, however, equal to a fee chargeable for at least 10 surface units.

The applicant must pay all costs of the granting of the concession. If the Ministry of Trade and Industry so directs, these costs must be paid to the

---

[2] Nielsen (2004), Country Report Finland, in: Department of Mining and Tunneling, lc.

Ministry in advance. The application may be rejected if the costs are not paid in accordance with the directions (Article 25 ML).

For the concession right, the concession holder, (unless he is the owner of the fields) must pay the land owner an annual fee, from the beginning of the calendar year following the year of issue of the permit certificate (Article 44 ML). The concession fee has, for each calendar year, to be paid by the 15th March of the following year at the latest.

12.3.2 Additional Legal Basics

Table 46: List of Acts significant for raw materials – Finland

Name of Law	No. of Law and Year of Issuing
Land Use and Building Act	Law No. 132 of 1999
Land Extraction Act	Law No. 555 of 1981
Environmental Protection	Law No. 86 of 2000, as amended by Law No. 137
Nature Conservation Act Law	Law No. 1096 of 1996
Environmental Impact Assessment Procedure	Law No. 468 of 1994
Act on the Land Information System and Related Information Service	2002
Act on the Right to Transfer State Real Estate Assets	Law No. 937 of 2002
Administrative Procedure Act	Law No. 434 of 2003
Code of Real Estate	1995
Value Added Tax Act	Law No. 1501 of 1993
Waste Tax Act	Law No. 495 of 1996 last amended by Law No. 1066 of 2002
Act on Compensation for Environmental Damage	1994
Act on Environmental Permit Authorities	2000
Act on Implementation of the Legislation on Environmental Protection	Law No. 113 of 2000
Act on Jointly Owned Forests	Law No. 109 of 2003

Name of Law	No. of Law and Year of Issuing
Act on Water Resources Management	2004
Forest Act	1996
Waste Act (1072/1993) last amended 2000	Law No. 1072 of 1993 last amended in 2000
Government Decision on the Safety of Construction Work	Law No. 629 of 1994 last amended by Law No. 702 of 2006
Regional Development Act	Law No. 602 of 2002

Environmental Protection Act

The objective of the Environmental Protection Act is to prevent the pollution of the environment and to repair and reduce damage caused by pollution; to safeguard a healthy, pleasant and ecologically diverse and sustainable environment; to prevent the generation and the harmful effects of waste; to improve and integrate assessment of the impact of activities that pollute the environment and to promote sustainable use of natural resources (Article 1). The following principles apply to (mining) activities that pose a risk of pollution (Article 4): harmful environmental impact shall be prevented or, when it cannot be prevented completely, reduced to a minimum; the best available technique shall be used; combinations of various methods, such as work methods, shall be used and such raw materials and fuels shall be selected as provide appropriate and cost-efficient means to prevent pollution.

Environmental Impact Assessment Procedure Act

The Environmental Impact Assessment Procedure Act and later amendments, together with the Ordinance on Environmental Impact Assessment Procedure No. 268 of 1999, regulate the extent, contents and administration of EIA analyses and reporting. The Environmental Impact Assessment Ordinance stipulates that in connection with exploitation of natural resources (minerals) must the following activities undertake EIA analyses: The extraction and processing of mining minerals if the annual extraction is at least 550.000 tonnes per annum. Quarries or sand and gravel pits if the excavation covers an area larger than 25 hectares, or if the extracted volume is at least 200.000 m^3 (solid or bank) per annum.

Nature Conservation Act

The aim of this Act is to maintain biological diversity; conserve nature's beauty and scenic value; promote the sustainable use of natural resources and the natural environment; promote awareness and general interest in nature; and promote scientific research (Article 1). Any action altering the natural surroundings is prohibited in a national park or strict nature reserve. Extraction of sand and stone materials and minerals, and any action that damages the soil or bedrock is thus prohibited in these areas (Article 13).

13 France

13.1 General Facts

The national territory covers 674 843 km² (Metropolitan France and Territory of the French Republic in the world) and the inhabitants are 65.073.482 (2009 estimation). The population density is 115/km². The GDP per capita for 2009 is estimated at $ 42.747. France is one of the founding members of the EU and has the largest land area of all members.

Constitutional Structure

The form of government is a unitary semi-presidential republic. France is divided into 26 administrative regions. The regions are further subdivided into 100 departments (i. e. 341 arrondissements; 4.032 cantons; 36.680 communes). Regions, departments and communes have an elected local government and an executive. Arrondissements and cantons are only administrative divisions.

13.2 Production of Raw Materials

Table 47: Production Data – France (World Mining Data, Weber and Zsak, 2008)

Resources		2002	2003	2004	2005	2006	Change 02/06	Change 05/06
Aluminium	(t)	463.200	444.100	446.900	437.900	442.100	−4,56	0,96
Bauxite	(t)	150.000	174.000	170.000	170.000	160.000	6,67	−5,88
Cadmium	(t)	154	120	120	100	100	−35,06	0,00
Gold	(kg)	1.724	1.744	0	0	0	−100,00	
Silver	(kg)	700	500	700	0	0	−100,00	
Baryte	(t)	65.000	50.000	75.000	82.000	30.000	−53,85	−63,41
Diatomite	(t)	68.000	75.000	75.000	75.000	75.000	10,29	0,00
Feldspar	(t)	700.000	720.000	670.000	650.000	650.000	−7,14	0,00
Fluorspar	(t)	105.000	110.000	105.000	110.000	80.000	−23,81	−27,27
Gypsum	(t)	3.700.000	4.500.000	3.500.000	3.500.000	4.800.000	29,73	37,14
Kaolin	(t)	335.000	340.000	350.000	370.000	300.000	−10,45	−18,92
Potash	(t)	145.000	80.000	70.000	68.000	0	−100,00	−100,00
Salt	(t)	4.205.000	3.900.000	4.500.000	4.900.000	5.200.000	23,66	6,12

Resources		2002	2003	2004	2005	2006	Change 02/06	Change 05/06
Sulphur	(t)	795.000	750.000	894.400	802.345	930.000	16,98	15,91
Talc	(t)	417.000	359.000	340.000	330.000	350.000	−16,07	6,06
Steam-Coal	(t)	1.920.000	2.234.000	872.000	780.000	452.000	−76,46	−42,05
Lignite	(t)	400.000	380.000	200.000	180.000	0	−1,0.00	−1,0.00
Nat. Gas	(Mm³)	1.400	1.300	1.200	1.100	1.200	−14,29	9,09
Oil	(t)	1.600.000	1.400.000	1.300.000	1.200.000	1.300.000	−18,75	8,33
Oil-Shale	(t)	14.000	13.000	12.500	12.200	12.300	−12,14	0,82
Uranium	(t)	24	11	8	8	0	−1,0.00	−1,0.00

Remarks:
Bauxit recoverd from dumps;
Steam-Coal inc. anthracite.

Table 48: Aggregates Production – Annual Statistics/France, 22 April 2008, quantities in million tonnes (UEPG 2008)

Sand & Gravel (1)		Crushed rock (2)		Marine Aggregates		Recycled Aggregates (3)		Manufactured Aggregates (4)	
2005	2006	2005	2006	2005	2006	2005	2006	2005	2006
170,0	167,0	223,0	233,0	n. a.	7,0	10,0	14,0	7,0	9,0

(1) Sand and Gravel: sold production including crushed gravel
(2) Crushed rock: sold production (excluding crushed gravel)
(3) Recycled Aggregates: materials coming from construction and demolition waste used in aggregates market
(4) Manufactured aggregates include blast-furnace-slag, electric-arc-furnace-slag, incinerator bottom ash (IBA), pulverised fuel ash (PFA)

13.3 Normative Basics

13.3.1 Primary Legal Basics

The primary legal basics of mineral extraction activity is the Mining Code ("Code Minier"; 1992, amended by Law No. 105 of 2004 and regulation No. 407 of 2006.

Besides that Law No. 93–3 of 1993 is important, it was modifying the legal regime and regulations for quarries. Prior to 1993, all minerals were regulated by the Code Minier and controlled by the Ministry of Industry. In 1976, quarrying activities were transferred to the category of "classified activities", which are controlled by separate legislation (Loi sur les Installations Classees). Although the category changed in 1976, a transitional period of 15 years followed before quarried activities were laced under the system for controlling classified activities (regulated by the Ministry of Environment). The Law No. 93–3 of 1993 (Loi No. 93–3 du janvier 1993 relative aux carriers) was implemented by decree in June 1994 (Decree No. 94–485 du 9 juin).[1]

13.3.1.1 General Rules

Table 49: Structure of the Code Minier, France

First Book: General Regime	Title I Classification of Mineral Deposits			Art. 1–6
	Title II Exploration of Mines			Art. 7–19
	Title III Exploitation of Mines			Art. 21–24
		Chapter I Concessions	Section I Conferment of concession	Art. 25–35
			Section II Report with the owners of the surface and third parties	Art. 36–44
			Section III Withdrawal and end of the concession	Art. 45 (Art. 46–49 abolished)
		Chapter II Permits of Exploitation of Mines		Art. 50–63
		Chapter III State owned Mines		Art. 64–67
		Chapter IV Particular Dispositions in Overseas Departments	Section I Approvals of Exploitation	Art. 68–68/8
			Section II Permits of Exploitation	Art. 68/9–68/18
			Section III Various Dispositions	Art. 68/19–68/20
			Section IV Exploration and Exploitation in the Sea	Art. 68/21–68-/4

---

1 Department of the Environment (1995), France Profile, lc.

Title IV Execution of Exploitation Tasks and Exploration of Mines	Chapter I Relations between the Explorers and Owners or with the Owners of the Surface		Art. 69–76
	Chapter II Financial Year of Administrative Surveillance and Measures to be taken in the Event of Accident		Art. 77–90
	Chapter III Termination of Mining Work and Prevention of Risks	Section I Termination of mining jobs	Art. 91–92
		Section II prevention and surveillance of mining risks	Art. 93–96
		Section III Agency for Prevention and Surveillance of Mining Risks	Art. 97
Title V Geothermal Issues			Art. 98–103
Title V a Underground Storage			Art. 104–104/8
Title VI Quarries (1)			Art. 105–119
Title VI a Collection of Titles of Researches and Exploitation and Renunciation of these Rights			Art 119/1–119/4
Title VI b Modification and Lease to Titles of Researches and Exploitation			Art 119/5–119/10
Title VII Transit in the Class of Mines of Substances previously subjected to the legal Regime regarding Quarries			Art 102–129
Title VII a Exploitation from Dumps and Waste of Quarries			Art 130
Title VIII Statements of Excavations and Geophysic Surveys			Art 131–137

	Title IX Expertises			Art 138–139
	Title X Establishment of Offences and Penalties			Art. 140–144/1
Second Book Particular regulations	Title I National Exploitation of Flammable Solid Minerals			Art. 145–171
	Title II State-owned Mines of Potash in Alsace and Organisation of the Potash Industry			Art 172–187 (abolished in 2006)
	Title III Certain Organisations or Special Regulations concerning Exploitation and Exploitation of Hydrocarbons	Chapter I, II, III		Art. 188–202 (abolished)
	Title IV Exploration and Exploitation of Hydrocarbons in Aquitaine			Art. 203–206 (abolished), 207
Third Book Social dispositions	Title I Working conditions and health and security at work	Chapter I Working conditions		Art 208–211/5
		Chapter II Health and security at work		Art 212–218/29
	Title II Miners Representative	Chapter I Base Miners Representative	Section 1 Functions	Art 219–224/6
			Section 2 Districts	Art 225–226/4
			Section 3 Elections	Art 227–241/10
			Section 4 Special dispositions	Art 242–250/1
		Chapter II Permanent delegates of the surface		Art 251–251/39
		Chapter III Municipality Dispositions		Art 252–252/12
	Title III Penal Dispositions			Art 253–260

According to Article 2 ML minerals are classified as follows:

- coal, lignite, or other fossil fuels, except peat, bitumen, hydrocarbon liquids or gases, graphite, diamond;
- salts of sodium and potassium in solid or dissolved form, except those contained in the brines used for therapeutic purposes;
- sulphate;
- bauxite, fluorite;
- iron, cobalt, nickel, chromium, manganese, vanadium, titanium, molybdenum, rhenium, tungsten;
- copper, lead, zinc, cadmium, germanium, indium, tin, scandium, cerium and other rare earth elements;
- niobium, tantalum;
- mercury, silver, gold, platinum, metals of platinum; helium, lithium, rubidium, caesium;
- radium, thorium, uranium or other radioactive elements.

Ownership of mineral rights

Minerals of high value or of national importance such as gold, silver, lead, copper, zinc, coal and lignite are defined as 'mined substances' in Article 2 ML ("Code Minier").

Both exploration and extraction rights to these are state-owned. Minerals not specified in Article 2 are classified as 'quarried substances' (Article 4), and include aggregates such as limestone, igneous rock, sand and gravel. Rights to extract quarried minerals belong to the owner of the land. Approximately 80% of quarried substances are aggregates, but other such minerals are limestone and clay.[2]

Minerals can, however, change categories. For example, following a public inquiry, a substance may be classified as 'mined' by a decree from the Council of State at any time (Article 5 ML).

Acquisition of mineral rights

Concession to mine Article 2 minerals are granted by an order from the Council of State (Conseil d'Etat), following consultation with the General Mines Council (Conseil General des Mines). This order specifies the extent and limits of the concession, the timespan of the mineral right, and the rights of the owners of the surface over the products of the mine.

---

2   Department of the Environment (1995), lc.

The establishment of a concession, regardless to the owner of the surface, creates a separate property right (Article 36 ML).

Whilst quarries are normally left at the disposal of the owners of the land, Article 109 of the Mining Code allows the public authorities to define areas in which authorisations for exploration and extraction permits may be issued without the consent of the owner of the land. Those powers may be used when the utilisation of a quarried material is sufficiently compromised to threaten regional and national balance.

13.3.1.2 Issuing of Permits

Exploration

Exploration for quarried minerals is judged on a case basis. There are no fees to be paid to the authorities and/or land owner during exploration.[3]

Exploration for "mined substances" may be undertaken (Article 7 ML) either by the owner of the surface or with his consent, after reporting to the Prefect; or, failing consent, with the authorization of the Minister for Mines (i. e. Minister of Industry). Exploration is an exclusive permit.

The exclusive exploration permit is granted by the competent authority, after competing, for 5 years (Article 9). This permit gives the holder the exclusive right to do all exploration work within the scope of it and dispose of products taken during research and testing that can entail. One can obtain an exclusive permit when he has the necessary technical and financial capacity to carry out the exploration work and meets the obligations mentioned in Articles 79 and 91. A decree of the Council of State defines the criteria for assessing these capabilities, criteria for allocation of securities and the procedure for examining applications for exploration permits. Applicants for exploration permits furthermore must provide a work programme. A note on the likely environmental impact of the exploration must also be submitted. Article 133 ML forces the operator of any geophysical or geochemical exploration, or of a study of heavy minerals to inform the Mine Inspectorate and to communicate the results.

At the request of its holder, the validity of a permit may be extended twice, each time in five years than in the same conditions as those laid down for its issue, except the competition. Each of these extensions is right, either for a period of at least three years. It is less than three years when the holder has fulfilled its obligations and supports the request to extend a financial commit-

---

[3] Department of the Environment (1995), lc.

ment at least equal to the agreed financial commitment for the previous period of validity, in proportion to the validity and size requested (Article 10).

Exploitation

Subject to the provisions of Article 22 ML, the mines can be operated only under a permit (Article 21). Applicants for an extraction permit do not need to own or be renting the land before making an application. However, they must have the authorisation of the owner to apply for an extraction permit. As mentioned: For mined materials of national importance, there is a procedure whereby the operator can apply for the temporary occupation of the land, in order to use the land required for mining, without obtaining the ownership of it (Article 109 ML).

According to Article 25 ML the concession is granted by an Order from the Council of State after a public inquiry related to competition and is also subject to the provisions of Article 26. The general and specific conditions are defined by decree from the Council of State and first made known to the petitioners. A person may obtain a mining concession if he has the technical capacity and financial resources to carry out mining operations and to meet the obligations mentioned in Articles 79, 79–1 and 91. A decree from the Conseil d'Etat defines the criteria for assessing these capabilities, criteria for allocation of securities and the procedure for examining applications for concessions.

During the term of an exclusive permit one can only obtain a concession, within the scope of this permit on the substances mentioned by the latter. The owner of an exclusive permit is entitled, if he so requests before the permit expires, to grant concessions on mineable deposits discovered within the scope of the permit during the validity of it (Article 26 ML). If an exclusive permit expires normally just before it was finally decided on an application submitted by the holder, the validity of the permit is extended by law without formality to the intervention of a decision concerning that request. This extension is only valid for substances within the boundaries set by the application concession.

The institution of concession involves the cancellation of an exclusive permit to the before mentioned substances and within the boundaries set by this concession. The right concerning an exclusive licensee to perform all research work within the scope of this permit is maintained.

The scope of a concession is determined by the concession. It is limited by the surface in depth and relying on a perimeter defined surface (Article 28).

The duration of mining concessions is determined by the concession (Article 29 ML). It cannot exceed 50 years. A mining concession can be subject to successive extensions, each of length less than or equal to 25 years. At the end of concession as provided by Decree from the Council of State: The deposit is returned to the free state, after completion of the work prescribed for the purposes of this Code; Dependency properties can be returned free of charge or sold to the State where the deposit remains usable, all the rights and obligations of the concessionaire is transferred to the State in case of loss or failure of the operation. Mining concessions imposed for an indefinite period expire December 31, 2018.

Withdrawal and end of the concession: Article 45 establishes the conditions under which the holder of a concession may waive, in whole or in part thereof.

Quarries

The departmental Prefect ('Préfet')[4] issues the extraction permit. The regional Directorates for industry, research and environment (DRIRE, 'Direction Régionale de l'Industrie, de la Recherche et de l'Environnement'[5]) play an important part in the permit process. The DRIRE examines the results of a public hearing and of the advice of the Municipality and of other organisations involved (e. g. water, military, etc.). The DRIRE reports to the departmental commission for quarries ('Commission Départementale de la nature, des paysages et des sites'), which writes an advice for the departmental Prefect. The report by the DRIRE is also forwarded to the public hearing commission, the mayor and the applicant. It is the Prefect who takes the final decision and publishes this decision. In most cases, the Prefect follows the advice by the commission for quarries. The permit, called a permit for temporary occupation, is normally limited to 30 years.[6]

The following information must be submitted, when applying for an extraction permit: maps and plans, an environmental impact study, a risk assessment, a plan for rehabilitation, and technical and financial guarantees. Applicants for an extraction permit do not need to own or be renting the land before making an application. However, they must have the authorisation of the owner to apply for an extraction permit.

According to Article 107 ML the quarries that have been authorized under Articles 3 and 5 of Law No. 76–663 of 1976 on classified installations for environmental protection or who have been regularly open under the mining code, is subject to the following provisions: If quarrying is likely to endanger

---

4   Central Government representative at local (departmental) level.
5   http://www.drire.gouv.fr/
6   Vervoort, A. (2004), Country Report France, in: Department of Mining and Tunnelling, lc.

safety and health of people, there is provided by the representative of the State Department the need for monitoring at the expense of the operator. Decrees identify measures of any kind, to improve safety or health conditions, to allow the execution of technical research required for these improvements and to ensure the correct use of the deposits. The competent authority (in enforcement under this code) can visit quarries at any time. It can also require documents of any kind. The owner of a quarry who has complied with the contract must compensate (pay) the owner if he continues its operations or sells the right to a third party.

Restoration

The rehabilitation of an extraction site forms part of the environmental impact study. As for other items, rules for the rehabilitation can be attached to the conditions of authorisation.

13.3.1.3 Authorities

The main responsible authority for mining is the Minister of Industry ("Conseil Général des Mines").

The National Government plays an overall regulatory role in the extraction of minerals. Responsibility for 'mined' substances lies with the Ministry of Industry, whilst the Ministry of Environment controls the exploitation of 'quarry' substances. Marine aggregates fall under the responsibility of the Ministry of Industry. For the mining of state-owned mineral resources (e. g. coal, lignite, metallic ores, etc.), itractional government delivers the extraction permit. The DRIRE, the regional Directorate for Industry, research and environment, evaluates applications for quarries, and it then reports to the general council for mining ("Conseil Général des Mines"). Finally, it is the Ministry of Industry which delivers a permit or refuses it.[7]

The monitoring of quarries and mines is stipulated in the permits. They cover a broad spectrum of issues, for example noise, dust, surface and groundwater. The measurements must be conducted or paid by the operator. The Prefects may impose new conditions for protecting the environment on the operator at any time during the operations, e. g. when the monitoring has shown a new risk or danger. Since the Law of 4 January 1993, the quarries are classed as classified installations regarding environmental protection (ICPE, "Instal-

---

7   Ibidem.

lations Classées pour la Protection de l'Environnement") and, hence, are initially supervised by the Ministry of the Environment.[8]

13.3.1.4 Fees and Taxes

There are no fees which must be submitted to either the government or the landowner during exploration (1994).

For "mined" substances, an extraction tax must be paid to departments and communes. In addition, a land tax is payable from the operator (1994). For mined materials, the operator must pay to obtain the rights to extract (after having obtained the extraction permit).

Regarding quarries: The permit holder is required to pay the owner of the surface a fee for having a base tonnage mined. Failing mutual agreement, the amount is fixed by the civil court at the request of either party, taking into account contracts for the licensing of quarrying similar to the consistency of cottage, value of materials being extracted (Article 115 ML).

A one-off tax or an annual tax is payable when authorisation for a quarry is granted, renewed or extended, as part of the legislation on installations classified for environmental protection purposes (ICPE, "Installations Classées pour la Protection de l'Environnement").

13.3.2 Additional Legal Basics

Table 50: List of Acts significant for raw materials – France

Name of Law	No. of the Law and Year of Issuing
Law, modifying the legal regime and regulations for quarries	Law No. 3 of 1993
Law on installations classified for environmental protection purposes (ICPE, 'Installations Classées pour la Protection de l'Environnement')	Law No. 663 of 1976
Law on prevention plans for mining risks	Law No. 245 of 1999
Regulation for the prevention and the surveillance of mining risks	Decree No. 353 of 2002
Environmental Law	Law No. 101 of 1995

---

8   Ibidem.

Name of Law	No. of the Law and Year of Issuing
Law on the prevention of technical and natural risks and the responsibility for the damages involved.	Law No. 699 of 2003
Law on the protection of countryside	Law No. 24 of 1993
Law regulating various aspects of water use and water protection	Law No. 3 of 1992
Civil Code	

14 Germany

14.1 General Facts

The national territory covers 357 022 km² and the inhabitants are 82,2 million. The population density is 230/km². The GDP per capita amounted to $ 40.874 (2009 estimation).

Constitutional Structure

The form of government is a democratic parliamentary state with 16 federal states. The country is divided into the national level, provincial level (16 federal states), district level (313 districts) and municipal level (12 320 municipalities).

14.2 Production of Raw Materials

Table 51: Production Data – Germany (World Mining Data, Weber and Zsak, 2008)

Resources		2002	2003	2004	2005	2006	Change 02/06	Change 05/06
Iron	(t)	58.711	60.300	57.700	37.796	43.680	−25,60	15,57
Aluminium	(t)	652.800	660.800	667.800	647.934	515.539	−21,03	−20,43
Cadmium	(t)	422	640	640	640	640	51,66	0,00
Baryte	(t)	100.993	109.506	93.624	88.591	85.524	−15,32	−3,46
Bentonite	(t)	495.310	478.796	404.549	352.374	393.998	−20,45	11,81
Feldspar	(t)	1.000.000	1.800.000	2.219.642	3.309.134	3.500.000	250,00	5,77
Fluorspar	(t)	34.429	33.289	33.203	35.364	53.009	53,97	49,90
Graphite	(t)	3.312	2.840	3.155	2.638	0	−100,00	−100,00
Gypsum	(t)	1.800.000	1.800.000	1.579.000	1.644.000	1.771.000	−1,61	7,73
Kaolin	(t)	3.681.953	3.600.000	3.752.000	3.768.000	3.815.000	3,61	1,25

14 Germany

Resources		2002	2003	2004	2005	2006	Change 02/06	Change 05/06
Potash	(t)	3.471.763	3.563.000	3.505.000	3.664.000	3.625.000	4,41	−1,06
Salt	(t)	14.695.000	16.308.000	18.696.000	18.731.000	19.146.000	30,29	2,22
Sulphur	(t)	1.093.091	1.014.133	976.729	1.054.800	1.113.802	1,89	5,59
Talc	(t)	12.000	9.000	6.000	5.000	4.000	−66,67	−20,00
Steam Coal	(t)	11.248.000	11.479.000	12.516.000	12.847.000	10.649.000	−5,33	−17,11
Coking Coal	(t)	17.961.000	17.274.000	16.635.000	15.171.000	13.113.000	−26,99	−13,57
Brown Coal	(t)	181.779.000	179.085.000	181.926.110	177.907.000	176.321.117	−3,00	−0,89
Natural Gas	(Mm³)	21.424	20.910	20.405	19.762	19.776	−7,69	0,07
Oil	(t)	3.768.054	3.808.946	3.516.318	3.572.462	3.515.401	−6,71	−1,60
Oil Shales	(t)	366.841	295.853	282.408	292.385	320.207	−12,71	9,52
Uranium	(t)	261	123	91	94	77	−70,50	−18,09

Table 52: Aggregates Production – Annual Statistics/Germany, 22 April 2008, quantities in million tonnes (UEPG 2008)

Sand & Gravel (1)		Crushed rock (2)		Marine Aggregates		Recycled Aggregates (3)		Manufactured Aggregates (4)	
2005	2006	2005	2006	2005	2006	2005	2006	2005	2006
263,0	277,0	174,0	270,0	n.a.	0,4	46,0	48,0	30,0	30,0

(1) Sand and Gravel: sold production including crushed gravel
(2) Crushed rock: sold production (excluding crushed gravel)
(3) Recycled Aggregates: materials coming from construction and demolition waste used in the aggregates market
(4) Manufactured aggregates include blast-furnace-slag, electric-arc-furnace-slag, incinerator bottom ash (IBA), pulverised fuel ash (PFA)

14.3 Normative Basics

14.3.1 Primary Legal Basics

The primary legal basics of mineral extraction activity is the Mining Law ("Bundesberggesetz") No. 1310 of 1980, as amended by Law No. 2585 of 2009.

However, there exists no uniform body of law on mineral extraction in Germany. "Free for mining minerals" (i. e. metallic ores, industrial minerals) and certain minerals owned by the land owner are covered by the Mining Law, whereas other minerals (i. e. construction minerals) are subject to other laws.

14.3.1.1 General Rules

Table 53: Structure of the Mining Law – Germany

Part 1: Preliminary Provisions	
	Articles 1–5
Part 2: Bergbauberechtigungen Mining licences	
Chapter 1: Free for mining minerals	Section 1: authorisation, permit, mining property Articles 6–23
	Section 2: Merger, division and exchange of mining property Articles 24–29
	Section 3: exploration and extraction royalty ("Feldes- und Förderabgabe") Articles 30–32
	Section 4: Notification of discovery Article 33
Chapter 2: Land owner minerals	Article 34
Chapter 3: "Zulegung"	Articles 35–38
Part 3: Exploration, exploitation and processing	
Chapter 1: General regulations for exploration and exploitation	Section 1: Exploration Articles 39–41
	Section 2: Exploitation Articles 42–47
	Section 3 Restraints and prohibitions Articles 48–49
Chapter 2: Notification, operation plan	Articles 50–57

Chapter 3: Responsible persons	Articles 58–62	
Chapter 4: Other provisions for the operation	Articles 63–64a	

Part 4: Authorisation for promulgation of mining regulations

Articles 65–68

Part 5: Mining supervision

Articles 69–74

Part 6: "Berechtsamtsbuch", "Berechtsamtskarte"

Articles 75–76

Part 7: Mining and property, public transport facilities

Chapter 1: Cession of property	Section 1: Legitimacy and requirements for cession of property Articles 77–83
	Section 2: Compensation Articles 84–90
	Section 3: Preliminary ruling, achievement and cancellation of the cession of property Articles 91–96
	Section 4: Premature Property Introduction Articles 97–102
	Section 5: Charges, compulsory execution, proceedings Articles 103–106
Chapter 2: Building restriction	Articles 107–109
Chapter 3: Mining damage	Section 1: Articles 110–113
	Section 2: liability for mining damages: Subsection 1: General regulations Articles 114–121 Subsection 2: Funds for mining damages Articles 122–123
	Section 3: Mining and public transport facilities Article 124
	Section 4: Observance of the surface Article 125

Part 8: Other activities and dispositions

Articles 126–131

Part 9: Special regulations for the continental shelf

Articles 132–137

Part 10: Federal test laboratory, commission of official experts, implementation	
Chapter 1: Federal test laboratory for mining	Articles 138–140
Chapter 2: Commission of official experts, implementation	Articles 141–143
Part 11: Legal process, fine and forfeit directives	
	Articles 144–148
Part 12: Transitional and final provisions	
Chapter 1: Former rights and contracts	Articles 149–162
Chapter 2: Closure and phase-out of mining labour unions	Articles 163–165
Chapter 3: Further transitional and final provisions	Articles 166–178

Figure 2: Laws concerning mineral extraction – Germany

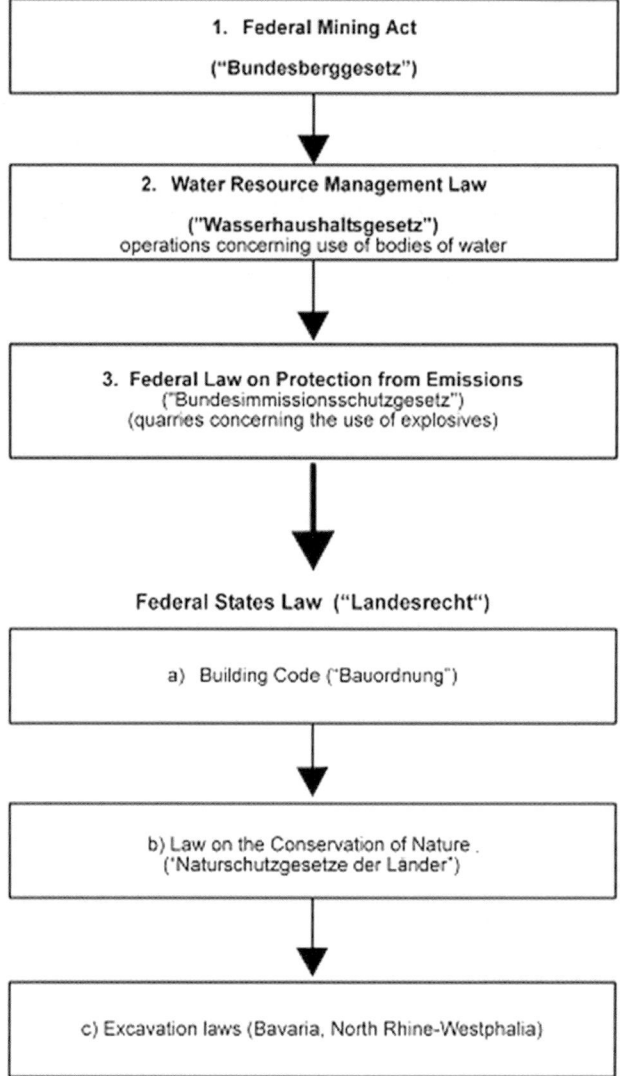

Basically the following approval procedures for mineral extraction can be distinguished between approval procedures under the Federal Mining Act ("Bundesberggesetz"), approval procedures under the Federal Law on Protection from Emissions ("Bundesimmissionsschutzgesetz"), approval procedures under the General Building Code ("Baugesetzbuch")/Federal Law on the Conservation of Nature ("Bundesnaturschutzgesetz") and approval procedures under the Federal Water Resources Management Act ("Wasserhaushaltsgesetz"), compare to figure 3.

Figure 3: Overview approval procedures – Germany (Müller, Schulz, 2000)

Land laws on nature conservation and Land building regulations	Federal Water Resources Management Act Land water law	Federal Mining Act		Federal Law on Protection from Emissions
↓	↓	↓	↓	↓
Dry extraction without blasts	Wet excavation, uncovering ground water	Free for mining minerals	Minerals owned by land owner and covered by mining law	Quarries where blasts are carried out; plants for crushing, grinding and grading
↓	↓	↓	↓	↓
Special case: Bavaria, N R-W Land excavation laws	Extension of waters Article 31 of the Federal Water Resources Management Act	Article 51 et sqq. of the Federal Mining Act (Operations plan approval procedure partly affected by concentration of jurisdiction)		Articles 4, 10, 19 of the Federal Law on Protection from Emissions
	↓			
	Plan approval plan adoption, here apply Art. 72–78 of the Administrative Procedure Act			

Classification of minerals

Mineral resources are divided into 3 categories (Article 3 ML):

- Free for mining minerals ("bergfreie Rohstoffe"), e. g. ores, graphite, salt, coal, oil and natural gas
- Minerals owned by the land owner and covered by the Federal Mining Act
- Minerals owned by the land owner and not covered by the Federal Mining Act, e.g sand, gravel, natural stone, gypsum, anhydrite, dolomite, marl, clay

The difference between the three categories is the way the land owner's disposal of the minerals located on his estate is regulated.

Ownership of mineral rights

a) "Free for mining minerals" are not owned by the land owner. The operator needs to acquire the rights to explore for and extract "free for mining" minerals by means of a mining permit.

b) Minerals owned by the land owner and covered by the Federal Mining Act (article 3 paragraph 4 Federal Mining Act)

The operator is entitled to extract them without requiring a mining permit. The entitlement arises from land ownership. Free for mining minerals as well as minerals owned by the land owner and covered by the Federal Mining Act are subject to regulations concerning the requirement of an operations plan and mining inspection. As far as exploration and extraction are concerned, "free for mining minerals" as well as minerals owned by the land owner and covered by the Federal Mining Act are treated equally (requiring operations plan as well as mining inspection). The difference between the two categories of minerals lies in the fact that the extraction permit of free for mining minerals must be granted, whereas the permit to extract minerals owned by the landowner and covered by the Federal Mining Act results from the ownership of land.

c) Minerals owned by the land owner and not covered by the Federal Mining Act

These minerals are not controlled by the mining law, but are directly subject to regulations under Civil Law ("Bürgerliches Gesetzbuch"). The land owner is entitled to the exploration and extraction of these minerals. For a considerable amount of these minerals there are no uniform permit requirements, but various national and federal states' regulations.

Acquiring mineral rights

Free for mining minerals: Under the Federal Mining Act, exploration and extraction of free for mining minerals require a mining permit, which represents merely a right granted by the State for the economic utilisation of free for mining minerals. In order to carry out mineral extraction operations it is furthermore necessary to obtain an approval of the operations plan from the competent authority.

Minerals owned by the landowner: In order to obtain mineral rights the mining company must be granted the right to use the land for the purpose of exploration and extraction of the minerals.

Table 54: Mining Permits for free for mining minerals – Germany (Müller, Schulz 2000)

	Authorisation	Permit	Mining Property
Contents	Exploration	Exploration and extraction	Exploration and extraction
Conditions	Article 11 ML	Article 11 (1) 6–10 ML Article 12 (1) 2 ML	Permit Article 13 ML
Terms			
Validity period	maximum 5 years	adequate term	adequate term
Extension	3 years each	admissible	admissible
Legal effect	priority over other approvals	protection like property	legitimate to burden property
Revocation			
Beginning	not within one year	not within 3 years	–
Break	longer than 1 year	longer than 3 years	longer than 10 years
Transfer	competent authority	competent authority	authority of approval

14.3.1.2 Issuing of Permits

Exploration

The authorisation ("Erlaubnis") confers the exclusive right to explore for minerals and to undertake activities associated with the exploration. An application in written form including a programme of work relating to the proposed technical execution of the measures and a time schedule must be submitted (Article 7 ML).

The permit ("Bewilligung") grants the approval holder the right to explore for and extract the minerals stated in the approval within a specified area (area of permit); furthermore the approval confers the entitlement to construct the necessary operating facilities and to demand a compulsory assignment of land. The most substantial reason for rejection is public interest (Article 8 ML).

Exploitation

The *mining property* ("Bergwerkseigentum") constitutes an even stronger and more extensive right than the permit (Article 9 ML). It can only be granted to a holder of an approval. With granting the mining property the permit expires. The mining property is granted on written application. The mining property comes into force with the delivery of the relevant document ("Berechtsamtsurkunde") to the applicant. Only when the decision on the granting has become final is the delivery admissible. The competent authority requests the land registry for registration of the mining property in the land register. The mining property is treated like a real property and can be encumbered with a mortgage and a charge on the land.

The mining permits alone do not provide entitlement for the operator to start with exploration and extraction activities, as they merely represent a title. Prior to carrying out these activities the license is required to ask for approval by submitting an operations plan which is being adapted to the ongoing operation.

Under the Federal Mining Act distinction is made between the following types of operation plans: Overall operations plan ("Rahmenbetriebsplan"), Main operations plan ("Hauptbetriebsplan"), Special operations plan ("Sonderbetriebsplan"), Collective operations plan ("Gemeinschaftlicher Betriebsplan"), Closing operations plan ("Abschlussbetriebsplan").

The *overall operations plan* represents the basis for subsequent operations plans and is usually approved for a period of 10 to 20 years. However, as it does not permit the installation and operation of a plant the operator is required to submit other operations plans in addition. The approval of the overall operations plan by the mining authority merely implies the ascertainment that the project meets the requirements necessary for approval.

The *main operations plan* is to be drawn up for the installation and operation of a plant. The contents of overall operations plans may vary, among others depending on extent and type of plant, methods of exploration and extraction, the respective branch, the phase of operation for which the main operations plan is drawn up.

The *main operations plan* is approved for a fixed term of 2 years. The authority is entitled to extend or shorten this period of time according to the respective circumstances. In order to carry on with the operation of a plant, it must be either drawn up again or extended. A main operation must comprise the following: details on the installations and facilities of the plant, development of the plant, methods of exploration and extraction used, working appliances used.

The competent authority may demand a *special operations plan* for certain parts of the operation or project. The plan is intended to complete and relieve the main operations plan. According to jurisdiction, special operations plans need not be limited to a fixed term. The number of special operations plans require increases with the extent of the mine.

Upon request of the relevant authority the *collective operations plan* is to be drawn up for workings and installations carried out, set up or operated by several plants under uniform aspects.

The mining company must submit the operations plan prior to the start of the planned operation to the competent authority. The mining authority involves other authorities and municipalities as planning authorities. Other authorities are to be involved if measures stated in the operations plan affect their area of responsibility. Upon obtaining the documents for application, the authorities comment on the respective case. The mining authority, however, is not bound to their comment and decides on its own on the approval of the operations plan. An extensive involvement of various authorities only occurs in the event of an overall operations plan. Once the involvement procedure and the assessment of approval requirements have been carried out, the authority either approves or rejects the operations plan in writing.

Drawing up and approval of the closing operations plan

In order to close a mining operation the operator must draw up a closing operations plan. The content of the plan should include an exact description of the technical operation and the period of time for which the closure of the operation is planned; proof of ensuring protection of a third party from dangers to life and health caused by the operation, also after the operation has ceased; proof of rehabilitation of the surface area affected by the operation.

The closing operations plan must be approved by the authority. The operator must carry out the approved closing operations plan. The extent of the measures to be taken results from the closing operations plan and its official approval which may incorporate incidental provisions for additional obligations.

14.3.1.3 Authorities

The main responsible authority for mining is the Minister of Economics.

The competent authority (regional mining authority) can arrange in accordance with Article 71 ML in individual cases, which measures are to be met for the execution of the regulations of this law. If a condition arises, which contradicts this law, to a direct danger for persons employed or third person, then the competent authority can arrange that the enterprise up to the establishment of proper conditions provisionally totally or partly is stopped.

14.3.1.4 Fees and Taxes

The permit holder of a right for exploration and extraction of minerals must pay a mining royalty to the Land in which the area of entitlement is located. The royalty for the exploration permit of free for mining minerals for industrial purpose ("Feldesabgabe") must be paid by the permit holder to the respective Land (Article 30 ML).

According to Article 31 the holder of the approval or the owner of mining property must pay a royalty for the free for mining minerals extracted within one year in the area of entitlement ("Förderabgabe").

14.3.2 Additional Legal Basics

Table 55: List of acts significant for raw materials – Germany

Name of Law	No. of Law and Year of Issuing
Federal Spatial Planning Act (Raumordnungsgesetz)	Law No. 2986 of 2008, as amended by Law No. 2585 of 2009, Article 9
General Building Code (Baugesetzbuch)	Law No. 2414 of 2004, as amended by Law No. 2585 of 2009, Article 4
Law for Environmental Impact Assessment (Gesetz über die Umweltverträglichkeitsprüfung)	Law No. 94 of 2010, as amended by Law No. 1163 of 2010, Article 11
Federal Water Resources Management Act (Wasserhaushaltsgesetz)	Law No. 2585 of 2009, as amended by Law No. 1163 of 2010, Article 12
Federal Forestry Law (Bundeswaldgesetz)	Law No. 1037 of 1975, as amended by Law No. 1950 of 2010
Federal Immission Control Act (Bundes-Immissionsschutzgesetz)	Law No. 3830 of 2002, as amended by Law No. 1163 of 2010, Article 3
Federal Nature Conservation Act (Bundesnaturschutzgesetz)	Law No. 2542 of 2009
Law for Soil Conservation (Bundes-Bodenschutzgesetz)	Law No. 502 of 1998, as amended by Law No. 3214 of 2009
Law for Trade and Industry Cycle/Waste (Kreislaufwirtschafts- und Abfallgesetz)	Law No. 2705 of 1994, as amended by Law No 1163 of 2010, Article 8
Administrative Procedure Act	
German Civil Code	
Environmental Information Act	1994

14.3 Normative Basics

Federal Law on Protection from Emissions

The purpose of the Federal Law on Protection from Emissions is to protect humans, animals and plants, soil, water, climate as well as cultural assets and other property. It also includes regulations relating to the provision against and prevention of harmful impacts on the environment. The procedure is related to the plant and not linked to the categories of minerals. Plants requiring approval have either to undergo a procedure involving the public or a simplified procedure, which means that public announcement or an appointment for discussion are not required. Even if a plant does not require an approval, the operator of the plant still has to take account of certain regulations.

Federal Water Resources Management

The Federal Water Resources Management Act regulates water conservation. The Act provides a framework, which is independently implemented and enacted in each land (individual land water laws). The scope of the Act comprises of coastal waters as well as ground water. Land ownership does not entitle to the use or the extension of waters, which is rather subject to a regulation of use under public law. The approval procedure under the Federal Water Resources Management Act is to be considered in the case of wet excavations of minerals not covered by the mining law. Utilisation under the Federal Water Resources Management Act must be rejected if the planned operation is likely to affect the good of the general public to an extent that cannot be balanced through ordinances.

Federal Nature Conservation

The purpose of the Federal Nature Conservation Act is to (Article 1) to maintain the efficiency of the balance of nature, to preserve the exploitability of nature's resources, to conserve fauna and flora, and to safeguard the variety, particularity and beauty of nature and landscapes, as a basis for mankind's existence and as a prerequisite to recreation in nature and in landscapes. According to Article 2 the pursuit of the objectives and of conservation of nature and of landscapes shall be guided by the following principles:

- Economical use of non renewable nature resources. Consumption of renewable resources shall be controlled in such a way as to ensure their continued and lasting availability.
- Any destruction of parts or components of landscapes in the course of mining for natural resources shall be avoided. Lasting damage to the balance of nature shall be prevented. Any adverse effects on nature and landscapes which inevitably result from the exploration and extraction of natural resources and from soil deposits shall be compensated for by recreating the original landscape or by relandscaping areas modelled after nature.

15 Greece

15.1 General Facts

The national territory covers 131 990 km² and the inhabitants (2009) are 11.247.285. The population density is 85,3/km². The GDP per capita for 2009 is estimated at $ 29.634.

Constitutional Structure

The form of government is a parliamentary republic. It consists of 13 administrative regions which are subdivided into 3 superprefectures and 51 prefectures (nomoi) there is also one autonomous area: Mount Athos.

15.2 Production of Raw Materials

Table 56: Production Data – Greece (World Mining Data, Weber and Zsak, 2008)

Resources		2002	2003	2004	2005	2006	Change 02/06	Change 05/06
Chromium	(t)	770	700	690	650	640	−16,88	−1,54
Nickel	(t)	19.229	18.000	18.116	23.000	18.224	−5,23	−20,77
Aluminium	(t)	165.262	167.797	166.634	165.300	164.528	−0,44	−0,47
Bauxite	(t)	2.468.865	2.442.312	2.396.065	2.495.100	2.284.380	−7,47	−8,45
Lead	(t)	28.036	2.000	0	3.000	12.400	−55,77	313,33
Zinc	(t)	35.000	3.000	0	4.000	18.000	−48,57	350,00
Silver	(kg)	74.800	7.000	0	2.300	25.500	−65,91	1.008,70
Baryte	(t)	10	20	25	50	0	−100,00	−100,00
Bentonite	(t)	1.056.598	1.156.642	1.030.556	1.200.000	1.166.575	10,41	−2,79
Feldspar	(t)	92.500	74.875	88.274	105.000	92.400	−0,11	−12,00
Gypsum	(t)	850.786	731.785	856.606	735.000	886.284	4,17	20,58
Kaolin	(t)	57.885	59.680	53.438	60.000	40.840	−29,45	−31,93
Magnesite	(t)	558.057	549.049	499.474	500.000	469.106	−15,94	−6,18
Perlite	(t)	838.997	1.079.036	1.053.388	1.000.000	1.039.150	23,86	3,92
Salt	(t)	126.118	192.161	187.522	190.000	169.700	34,56	−10,68
Talc	(t)	670	500	200	500	250	−62,69	−50,00
Lignite	(t)	71.074.009	69.410.756	71.237.228	71.200.000	63.362.562	−10,85	−11,01

Resources		2002	2003	2004	2005	2006	Change 02/06	Change 05/06
Nat. Gas	(Mm³)	36	27	23	30	23	−36,11	−23,33
Oil	(t)	186.661	136.795	132.890	172.000	94.391	−49,43	−45,12

Table 57: Production of aggregates – Greece (Metric tons unless otherwise specified), (USGS, 2008)

		2002	2003	2004	2005	2006
Silica		125.000	130.000	125.000	125.000	125.000
Stone:						
Dolomite		90.000	90.000	40.000	60.000	60.000
Marble	m³	178.839	233.436	144.860	151.180	150.000
Flysch		80.000	75.000	158.887	93.509	95.000
Quartz, microcrystalline	1.000 metric tons	150	150	150	150	150

15.3 Normative Basics

15.3.1 Primary Legal Basics

There is no single Law or Mining Code. Most important of all is Legislative Decree No. 210 of 1973 commonly referred to as "The Mining Code" as amended by Law No. 274 of 1976. This comprises almost the entire mining law.[1]

All provisions concerning the exploitation of industrial minerals and aggregates are defined by the Regulations on Mining and Quarrying Activities (KMLE). This was initially established as a Ministerial Decision in 1984 and was later amended by the Joint Ministerial Decision 14 080 in 1996.

---

[1] Agioutantis, Z. (2004), Country Report Greece, in: Department of Mining and Tunnelling, lc.

15.3.1.1 General Rules

"The Mining Code" distinguishes three categories, for the purpose of controlling exploration and exploitation:[2]

(i) metallic ores;
(ii) industrial minerals and marbles;
(iii) aggregates.

Legislative Decree 4433/1964 as amended by Law 273/1976 defines the rights of the State to explore for mineral ores on both its own behalf and for private enterprises. Law 669/1977 and the related Laws 1428/1984 and Law 2115/1993 concern the exploration and exploitation processes of quarried minerals (primarily marbles, industrial minerals and aggregates).

Mineral rights

As described above, the Mining Code differentiates between mineral ores and quarrying minerals, and in doing so, between the rights of mineral ownership and land ownership. For mineral ores, the provisions of the Mining Code establish the freedom to mine by the private sector, with the exception of those minerals whose rights of exploitation belong to the State. Based on the principle of free mining, the right of investigation and consequent exploitation is granted to the first applicant. The Mining Code establishes exclusive rights to mine-owners to explore and exploit all ore minerals under concession, except when this is not in the national interest (e.g. for environmental reasons). Furthermore, mine owners are given the legal right to safeguard future mine reserves. In contrast to mineral ores, landowners are granted the right to exploit quarry minerals (i.e. marble, industrial minerals and aggregates). Law 1428/1984 defines that exploitation of aggregates can only take place in the quarrying areas approved by the State for their suitability (in terms of their quality and their environmental impact).

The ownership rights and the rights of exploration, research and exploitation for the following minerals belongs to, and are taken up exclusively by, the State (Mining Code, Articles 3 and 6). In addition, the Government owns the rights to metallic minerals (including some non-metallic strategically important minerals such as feldspar), but the rights of exploration and exploitation can be granted by the State to private operators. Marine sand and gravel rights belong to the local prefecture and the Ministry for the Environment, Energy and Climate Change former Ministry of Development. Other rights of exploration and exploitation rest with the land owner.

---

2 Ibidem.

Obtaining mineral rights is governed by the Mining Code (Articles 3–14). Also Articles 147–156 refer to mineral rights that belong to the government (i. e. for fossil fuels, etc) and how these are administered.

15.3.1.2 Issuing of Permits

Exploration

The procedures to obtain permit for exploration activities are detailed in the Mining Code as amended by subsequent laws and environmental restrictions.

There are two different procedures to obtain permit for exploration activities. One pertains to metallic minerals and one to industrial minerals and aggregates. For the metallic minerals one has to obtain a Preliminary Exploration Permit (Articles 20–43 Mining Code) or exploration concession. This is valid for 2 years. If successful, it should be followed by an exploitation concession (Articles 44–64 Mining Code), which is valid for 50 years renewable. The concession is granted by the Government. In the case of industrial minerals and aggregates one should obtain an exploration permit (either for private or public lands). The latter is governed by Law 669 as amended by Laws 1428 and 2115 and taking into account environmental restrictions in Law 1650 and others.

Procedures for obtaining permit for exploration activity: The law differentiates between permit for exploration activity on metallic minerals and industrial minerals/aggregates. In the first case, procedures are very rigorous and are detailed in the Mining Code (Articles 15–43).

In the second case, if exploration is to be performed on private land, permit from the owner is sufficient (Law 669, article 10, as amended by Article 21 of Law 2115). Such activities should be completed within a year renewable to one more year. If exploration is to be performed on public lands, then permit from the prefect should be obtained. A reply to the petition should be issued within 6 months (Law 669, article 10, as amended by article 21 of Law 2115) provided all the paperwork is in order.

Extraction

An exploitation concession (Articles 44–64 Mining Code), which is valid for 50 years renewable is granted for metallic minerals. The concession is granted by the Government. In the case of industrial minerals and aggregates one should obtain an exploitation permit (either for private or public lands).

The latter is governed by Law 669 as amended by Laws 1428 and 2115 and taking into account environmental restrictions in Law 1650 and others. For industrial minerals and aggregates the exploitation permit may be valid for a number of years depending on the time that it was originally issued. For example there are cases where the permit can be renewed up to 40 years. The permit is granted by the Government.

There is a large number of Ministerial Decisions, Presidential Decrees and other legal paperwork amending or clarifying aspects of the exploitation of industrial minerals and aggregates, especially with respect to environmental controls, fees and fines, etc. This has been coded by different agencies (i. e. the technical chamber of Greece, the Inspectorates of Mines of Northern and of Southern Greece, etc). This information is available on line (in Greek).

Restoration

The restoration is a responsibility of the operator, regardless of the minerals rights ownership. There are certain bonds required in legislation for aggregates and industrial minerals, but not for metallic minerals. There are also bonds with no legal requirements but frequently used in order to enforce the EIAs. The state and landowner have the ability to impose restoration and aftercare conditions. There is a provision to allow local authorities to manage such lands.

15.3.1.3 Authorities

The main responsible authority for mining is the Ministry for the Environment, Energy and Climate Change former Ministry of Development. According to this Ministry, responsibility for authorizing the exploration and exploitation permits is shown in the table below:

Table 58: Responsibility for issuing mineral permits

Mineral Category	Level of authority issuing the permission	
	Exploration	Exploitation
I. Mining Ores	Prefectural	National
II. Industrial Minerals & Marbles	Prefectural	National (for industrial minerals)
		Interprefectural (for marbles)
III. Aggregates	Non existing (i. e. reserves are plentiful)	Prefectural

In practice, the controls of minerals developments is much more complicated and fragmented. Because of the nature of the permit system in force, a number of government ministries and agencies are involved in the process. Those with particular responsibilities are described below:

The Ministry for the Environment, Energy and Climate Change is responsible for ensuring that mineral extraction is in line with the economic interests of the country. The Institute of Geology and Mineral Exploration (IGME) advises the Ministry on the technical aspects.

Monitoring and various degrees and instruments of enforcement are specified in Law 1650/1986, as amended by Law 3010/2002. There is a national and regional responsibility level for monitoring and enforcement and the enforcement powers include fines, revocation of permit, injunction and imprisonment.

15.3.1.4 Fees and Taxes

Exploration

The administrative fees for application are the same for cases where the mineral rights are owned by the State and the Landowner. The fee is 190 € for the application plus 2641 € for areas smaller than 5 km² and 190 € plus 3815 € for areas greater than 5 km². For industrial minerals and marbles the administrative fee for the application is 1467 €. There are no other payments made to the Government or to the Landowner.[3]

Extraction

In this case there are no administrative fees for application but there are other payments made to the Government. So, if the mineral rights are owned by the State, there are royalties (1%–10%) paid to the State. This is true in the case the State leases the deposit. It is not valid when the State issues an exploration permit concession and a deposit is found. In the case where the mineral rights are owned by the landowner there are payments made to the landowner which vary according to private deals.[4]

---

3 Ibidem.
4 Ibidem.

15.3.2 Additional Legal Basics

Table 59: List of acts significant for raw materials – Greece[5]

Name of Law	No. of the Law and Year of Issuing
Legislative Decree 4433, Exploration by the state and other mining issues	4433 of 1964
Law 272, Establishment of the Greek Institute of Geology and Mineral Exploration (I. G. M. E.)	272 of 1976
Law 273, Amendment of Legislative Decree 4433 of 1964	273 of 1976
Law 274, Amendment of Mining Code (L. D. 210/1973)	274 of 1976
Law 669, Exploitation of industrial minerals and marble	669 of 1977, later amended ly Law 2115 of 1993
Law 998, Protection of forest lands concerning mining activity (later complemented by Joint Ministerial Decision 183037/80)	998 of 1979
Presidential Decree 1180, Regarding pollution maximum allowable limits on gases, liquids, solids and noise	1180 of 1981
Law 1428, Exploitation of quarries, and aggregate materials	1428 of 1984, later amended by Law 2115 of 1993
Ministerial Decision for the Regulation on mining and quarrying activities (K. M. L. E.)	1984
Law 1650, Protection of the environment (later amended by legislative decrees and ministerial decisions)	1650 of 1986
Presidential Decree 126, Administration of forest areas	126 of 1986
Law 1739, Management of water resources	1739 of 1987
Ministerial Decision 5813, Permitting of development of water resources by private bodies	5813 of 1989
Joint Ministerial Decision 69 269/5387, Environmental impact assessment studies	69 269/5 387 of 1990
This amends articles 3 and 4 of Law 1650 based on EC Directive 85/337	

---

5   Ibidem.

15.3 Normative Basics

Name of Law	No. of the Law and Year of Issuing
Joint Ministerial Decision 75 308/5512, Environmental assessment studies, amending article 5 of Law 1650	75 308/5 512 of 1990
Ministerial Decision 6812, Extension of permitting for quarrying	6812 of 1993
Ministerial Document 17, Regarding Joint Ministerial Decision 69 269/5387	17 of 1994
Joint Ministerial Decision 95 209, Environmental impact assessment administrative issues	95 209 of 1994
Joint Ministerial Decision 84 229, Environmental impact assessment for specific public works	84 229 of 1996
Ministerial Document 9, Pre-approval of siting	9 of 1996
Joint Ministerial Decision 14 080, Minimum specifications for health and safety of miners (in accordance to European Directive 92/104	14 080 of 1996
Joint Ministerial Decision 114 218, Specifications and management of solid wastes	114 218 of 1997
Presidential Decree 256, Amendment to Presidential Decree 541 of 1978 for additional environmental assessment study categories	256 of 1998
Land planning and sustainable development	2742 of 1999
Harmonization of Law 1650 with European Directives 97/11 and 96/61. Water management issues	3010 of 2002
Joint Ministerial Decision 11 014, Amendment of Laws 1650 and 3010 for environmental impact assessment studies	11 014 of 2003
Joint Ministerial Decision 1726, Environmental Impact and assessment for permitting of power generating stations using renewable resources	1726 of 2003
Protection of forests' ecosystems	3208 of 2003

16 Hungary

16.1 General Facts

The national territory covers 93,030 km² and the inhabitants are 10,1 million. The population density is 110/km². The GDP per capita amounts to about € 12.926 (2009).

Constitutional Structure

The form of government is a republic. It is divided into a national level, regional level (7 regions), provincial level (19 provinces) and municipal level.

16.2 Production of Raw Materials

Table 60: Production Data – Hungary (World Mining Data, Weber and Zsak, 2008)

Resources		2002	2003	2004	2005	2006	Change 02/06	Change 05/06
Manganese	(t)	11.400	11.500	13.100	13.700	14.000	22,81	2,19
Aluminium	(t)	35.300	35.000	34.300	31.000	30.000	–15,01	–3,23
Bauxite	(t)	294.000	300.000	304.000	304.000	507.400	72,59	66,91
Gallium	(t)	5	6	6	0	0	–100,00	
Bentonite	(t)	4.000	5.300	9.300	4.900	4.300	7,50	–12,24
Feldspar	(t)	45.000	40.000	35.000	33.000	30.000	–33,33	–9,09
Gypsum	(t)	108.000	62.000	55.000	35.000	30.000	–72,22	–14,29
Kaolin	(t)	1.000	1.300	300	400	450	–55,00	12,50
Perlite	(t)	72.000	60.000	65.000	69.900	140.000	94,44	100,29
Talc	(t)	900	800	700	650	500	–44,44	–23,08
Steam Coal	(t)	660.000	667.000	260.000	240.000	100.000	–84,85	–58,33
Brown Coal	(t)	12.100.000	12.692.000	10.965.000	9.602.000	9.952.000	–17,75	3,65
Natural Gas	(Mm³)	3.100	3.100	3.285	3.159	3.100	0,00	–1,87
Oil	(t)	1.050.000	1.100.000	1.077.000	948.000	1.300.000	23,81	37,13

Table 61: Production of aggregates – Hungary (Metric tons) (USGS, 2008)

		2002	2003	2004	2005	2006
Sand and gravel:						
Gravel	1.000 metric tons	29.138	35.000	33.544	33.500	34.483
Sand:						
Common	1.000 metric tons	12.000	12.000	12.500	12.800	11.634
Foundry		152.000	162.600	138.200	138.000	120.000
Glass		317.000	225.300	163.900	164.000	251.000
Stone:						
Dimension, all types	1.000 metric tons	5.626	5.500	5.000	5.000	5.000
Dolomite	1.000 metric tons	4.196	4.398	7.200	7.200	7.933
Limestone	1.000 metric tons	7.152	2.459	3.014	3.014	3.257

16.3 Normative Basics

16.3.1 Primary Legal Basics

The primary legal basics of the mineral extraction activity is the Mining Law No. XLVIII of 1993 as amended by Law No. CXXXIII of 2007.

16.3.1.1 General Rules

Table 62: Structure of the Mining Law – Hungary

Part 1: General Provisions	Article 1: Effect of the Act
	Article 2: Exercise of Activities
	Article 3: Right of the State
	Article 4: Surface Pre-Prospecting that May Be Performed on the Basis of a Report
	Articles 5–7: Prospecting and Exploitation that may be performed on the Basis of the Permit of the Authorities (Liberalised Activities)

Part 2: Concession	Article 8: Concession
	Article 9: Designation of Concession Areas
	Article 10: Concession Tender
	Article 11: Assessment of Bids
	Article 12: The Concession Contract
	Article 13: The Concession Company
	Articles 14-15: Mining Concessions Prospecting, Exploration and Exploitation
	Article 16: Conveyance and Underground Storage of Hydrocarbons
	Article 17: Prospecting for Geothermal Energy, and Exploitation thereof for the Purpose of Generating Energy
	Article 18: Transfer of the Concession
	Article 19: Exercise of Mining Activity under Concession
Part 3: General Rules of the Performance of Mining Activities	Article 20 Mining Royalty
	Article 21: Licensing of Mining Activities
	Article 22: Licensing of Prospecting
	Article 23: Licensing of Exploration, Exploitation and of the Utilisation of Waste Stockpiles
	Article 24: Construction and Operation of Hydrocarbon Conveying Pipelines
	Article 25: Geological Information Supply and Handling
	Article 26: Mine Plot
	Article 27: Technological Operation Plan
	Article 28: Rules of Operation
	Article 29: Joint Exploitation
	Article 30: Suspension of Exploitation
	Article 31: Rules of Construction
	Article 32: Security Zone and Protective Pillars
	Article 33: Mine Plan
	Article 34: Safety of Mining, and Plant Supervision of Activity
	Article 35: Serious Accidents and Serious Breakdowns
	Article 36: Landscape Rehabilitation
	Article 37: Mine Damages
	Article 38: Limitation of Surface Real Estate Property
	Article 39: Construction Ban and Restrictions
	Article 40: Right to Use of Water
	Article 41: Fines, Measures and Guarantees
	Article 42: Closing of Mines, Abandonment of Fields

Part 4: State Supervision of Mining	Articles 43–44: Mine Supervision
	Article 45: Supervision by the Mine Supervision of the Raising to the Surface of Abyssal Waters
	Article 46: Authority Supervision of Underground Storage Spaces
	Article 47: Procedural Fee
	Article 48: Hungarian Geological Survey
Part 5: Definition of Terms	Article 49
Part 6: Moral and Financial Appreciation of Miners	Article 49A
Part 7: Provisions of Putting into Force, and Temporary Provisions	Article 50

Ownership of mineral rights

As occurring mineral raw materials are the property of the state. The state has the responsibility about issuing mineral rights. Exploited mineral raw materials are transferred to the property of mining entrepreneurs. The Minister of National Development can for a definite period of time, relinquish the exercise of the activities listed in Article 8 by concession contracts (Article 3 ML).

16.3.1.2 Issuing of Permits

Prospecting and exploration

Article 4 ML regulates prospecting which can be performed on the basis of an agreement concluded with the owner of the real estate, and of a preliminary report communicating the date of the commencement of prospecting to the mine supervision (Article 4 (1) ML).

Articles 5–7 and Articles 22–23 ML regulate the application requirements related to exploration and extraction (without tender). The competent authority issues the exploration and exploitation permits in related mining areas with the involvement of other co-authorities (Environmental Inspectorate,

National Park Directorate, Directorates of Water Management and other authorities (Article 5 ML).

The exploration permit grants the operator an exclusive right as well as of initiating the establishment of the status of mine plot within the deadline defined in the permit (Article 6 ML).

Exploitation phase

Articles 8–19 ML regulate the issuing of concession of selected raw materials. In defined concession areas mining only can be conducted on basis of a concession contract. By a concession contract concluded with domestic or foreign legal entities and natural persons, the competent authority can for a definite period of time, relinquish the following: a) prospecting for, exploration and exploitation of raw materials (Article 8 ML).

The Minister of National Development may designate concession areas for an open tender. After evaluation the bids by a ministerial panel, the winner and the Minister conclude to a concession contract in which they agree in a work programme and the guarantees of good performance.[1]

Based on the available geological details available the competent authority takes stock of the closed areas that may be designated for concession (Art 9 (1) ML). In the invitation to tender, the competent authority indicates the closed areas where, based on the analysis of the economic, environmental, natural and social effects and with regard to the existing natural resources, the mining of raw materials appears to be promising (3).

Article 10 regulates the public tender. The competent authority invites a public tender for conclusion of the concession contract. In addition to the contents defined in Act XVI of 1991 on Concession, the invitation to tender has to contain:
a) definition of the confines of area, in respect of which the tender has been invited, indicating whether third parties have already obtained the right of mining raw materials;
b) definition of the raw materials, for the mining of which the concession has been tendered;
c) environmental, land, water, health care, nature and landscape protection requirements of the activity under concession, and the obligations concerning the provisions of guarantee for meeting the requirements;

---

[1] Hamor, T. (2004), Country Report Hungary, in: Department of Mining and Tunnelling, lc.

d) requirements concerning the contents of the work programme to be submitted;
e) conditions of participation in the invitation to tender (e. g. entrance fee, information on the economic and financial situation of bidder);
g) conditions of landscape recultivation in the area affected by mining activities;

Article 12 ML regulates the concession contract. The competent authority concludes a concession contract with the awarded bidder. Concession contracts can be concluded for a maximum period of 35 years, which may be extended on one occasion, at the most, by half the period of the concession contract (Article 12 (1) ML).

The planned period of exploration within the period of the concession can not be longer than 4 years, and can, on one occasion, be extended by half at the most. The operator within 1 year of completion of exploration can apply for the area to be qualified as a mine plot (Article 14 ML).

If the operator of the concession fails to commence exploitation within the deadline defined in the contract, but within 5 years of qualification of the area as a mine plot, at the latest, he must pay the compensation defined in the contract for the recovering the mining royalty lost. If it fails to meet the obligation of paying the above compensation, the concession is terminated.

Whilst in the case of concession contracts the consent of the Minister of National Development is required, in the case according to Articles 5–7 ML the consent of the mining authority is sufficient (Articles 18, 26 of the Mining Act).

According to Article 30 ML exploitation can be suspended for 3 years. In the case of the suspension of exploitation for a period of more than 3 years, the mine authority can initiate closure of the mine and performance of landscape rehabilitation.

The approval of the Minister is required for transfer to another party of exercise of the mining activity under concession (Article 18 ML).

General rules of the performance of mining activities

The mine supervision has to be licensed to explore for certain mineral raw materials – within the framework of the concession, by approving the work programme, – in other cases, by issuing the prospecting licence based on the technological operation plan (Article 27 ML).

According to Article 22 (3) ML exploration permit provides the operator an exclusive right of explore for raw materials, and, based on successful exploration, of having the status of mine plot established.

In an exempted location, exploration will be licensed by the mine supervision with the approval of the special authorities, and in agreement with the party concerned. If exploration affects a nature protection area, the operator has first to obtain the permit of the nature protection authority (Article 22 ML).

Besides issuing of mineral rights, the operator needs the permit of an operation plan by the competent authority. The operation plan shall be approved by the mine supervision. The operation plan must be prepared with regard to the technical, safety, health protection and fire regulations, as well as for the fulfilment of the technical plan of landscape rehabilitation (Article 27 ML).

Closing phase

Closing of mines is regulated in Article 42 ML. The operator has to prepare a technical plan. For closing the mine, the possibility of using the underground workings of the closed mine for other purposes shall also be examined (1). The underground workings that will not be further used must be abandoned in such a condition that it should not be a hazard to the environment, or the surface (Article 42 (2) ML).

16.3.1.3 Authorities

The main responsible authority for mining is the Minister of National Development (Hungarian Office for Mining and Geology).

According to Article 42 (4) ML: With the exception of the cases defined in legal rule, the authority type matters falling under the competence of the mine supervision, the mine inspectorate (mining captainship) competent in the region must proceed at the first instance, and the Hungarian Office for Mining and Geology has to proceed at the second instance.

The Hungarian Office for Mining and Geology is an organ of the state administration, with national competence, operating under the control of the Government, the supervision of which is exercised by the Minister, who also exercises the employer's rights in respect of the President of the Office (5). The headquarters and region of competence of the regional mine inspectorates will be defined by the Minister, on the proposal of the President of the Hungarian Office for Mining and Geology (6).

According to Article 43 ML the specialised administrative tasks of the state related to mining have to be performed by the mine supervision. In the course of the performance of the activities falling under its supervision (Articles 44 to 46), it is the task of the mine supervision to protect the life, physical condition and health of the workers, and to control compliance with the rules regarding the management of mineral resources, the protection of the environment, landscape and nature, technical safety and fire protection.

16.3.1.4 Fees and Taxes

According to Article 20 (1) ML the state is entitled to profit sharing, to a mining royalty, for the raw material exploited by the operator, who acquires the property thereof. The rate of the mining royalty is the following, with regard to the value of the quantity of mineral raw material exploited: 5% in the case of non-metallic mineral raw materials exploited by open-pit mining, with the exception of energy resources; 2% in the case of other solid raw materials (2).

According to Article 20 (4) ML mining activity performed on the basis of a concession contract (Article 12), the rate of mining royalty will be defined by the competent authority in accordance with the different types of mineral raw materials, natural features affecting success of exploitation, – and other public interests, in respect of each location of exploitation.

The amount of the mining royalty expressed in cash is the product of the rate of the mining royalty defined in %, and the value of the mineral raw materials exploited. The rules concerning the calculation of the value of the raw materials exploited will be established by the Minister, in agreement with the Minister of Finance (Article 20 (6) ML).

A segregated fund is formed from 10% of the mining royalty paid annually, for funding the landscape rehabilitation tasks (not performed) that may not be assigned to mining entrepreneurs. The rules of the utilisation of the fund are established by the Government in a Decree (Article 20 (8) ML).

16.3.2 Additional Legal Basics

Table 63: List of Acts significant for raw materials – Hungary

Name of Law	No. of Law and Year of Issuing
Law on the general rules of environmental protection	Law No. LIII of 1995 as amended by Law No. LV of 2001
Law on the water management	Law No. LVII of 1995
Act on concessions	Law No. XVI of 1991
Act on regional development and regional planning	Law No. XXI of 1996
Decree on activities that affect the quality of groundwater	No. 33 of 2000
Act of Land	1987
Act LVII Competition Act 1996	1996
Act LV on arable land	1994
Government Decree on activities that affect the quality of groundwater	No. 33 of 2000(III.17.)
Joint Decree KöM-EüM-FVM-KHVM of the ministers of environmental protection, public health, agriculture and regional development, and of traffic, communication and water management on the limit values necessary to protect the quality of groundwater and the geologic medium	No. 10 of 2000 (VI.2.)
Civil Code Act	Act IV 1959
Act CXVI on atomic energy	1996
Act IL on bankruptcy proceedings, liquidation proceedings and voluntary dissolution	1991
Act LXXVIII on the formation and protection of the built environment	1997
Act XCIII on labor safety, consolidated with MüM Decree No. 5/1993 (XII. 26.) of the Ministry of Labor	1993

Act LIII of 1995 on the general rules of environmental protection

The objective of the Act is to develop a harmonious relationship between humans and their environment, to protect the components and processes of the environment and to provide for the environmental conditions of sustainable development (Article 1). The extent of exploitation, the extent of the impact on the environment arising when the tailings produced in connection with mining, and the dressing and processing of mining products are disposed of, as well as the impact arising as a result of other activities linked to mining activities, may not exceed the standards established in a legal rule, or the decision of an authority made in accordance with the provisions of a legal rule (Article 17).

Nature Conservation Act

The scope of the Act LIII of 1996 on nature conservation covers all natural resources, natural environment, natural values and other elements of the environment which are essential for nature. In this sense mineral resources are within the scope.

Waste Management Act

The Act XLIII of 2000 on waste management applies to any waste and waste management activities and installations.

17 Iceland

17.1 General Facts

The national territory covers 103 001 km² and the inhabitants are 317.593. The population density is 3,1/km². The GDP per capita is estimated at $ 37.976 (2009).

Constitutional Structure

The form of government is a parliamentary republic. Iceland is divided into regions (8) constituencies (6), counties (23), and municipalities (79).

17.2 Production of Raw Materials

Table 64: Production Data – Iceland (World Mining Data, Weber and Zsak)

Resources		2002	2003	2004	2005	2006	Change 02/06	Change 05/06
Aluminium	(t)	263.528	265.900	270.600	273.318	325.000	23,33	18,91
Diatomite	(t)	26.494	30.000	19.332	3.236	3.200	−87,92	−1,11
Salt	(t)	4.500	4.500	4.600	4.600	4.700	4,44	2,17

Remarks:
Aluminium production from imported resources; no indigenous mine production

Table 65: Production of aggregates – Iceland (USGS, 2008)

		2002	2003	2004	2005	2006
Sand:						
Basaltic	m³	1.200	1.200	1.300	1.300	1.200
Calcareous, shell	m³	80.000	80.000	80.000	80.000	75.000
Sand and gravel	1000 m³	4.200	4.200	4.200	4.300	4.200
Silica dust		22.579	23.830	22.533	22.992	24.955
Stone, crushed:						
Basaltic		95.000	96.000	96.000	97.000	95.000
Rhyolite	m³	18.000	18.000	19.000	19.000	18.000

17.3 Normative Basics

17.3.1 Primary Legal Basics

The primary legal basics of mineral extraction activity is Nature Conservation Law No. 44 of 1999.

17.3.1.1 General Rules
17.3.1.2 Issuing of Permits

The provisions of Chapter VI Nature Conservation Law (NL) regarding "extraction of materials from the earth" applies to extraction of materials from the ground, the beds of watercourses and, as applicable, extraction from or under the seabed within Iceland's territorial waters and exclusive economic zone (Article 45 NL).

The planning of *extraction areas* is subject to the provisions of the Planning and Building Act, No. 73/1997 (Article 46 NL). Furthermore authorisation for extraction is subject to the provisions of the Act on Icelandic National Ownership of Seabed Resources, No. 73/1990. The Minister of Industry must seek, however, the opinion of the Nature Conservation Agency before granting permit (Article 47 NL).

All extraction on land and from or under the seabed is subject to the operating permit of the local authority concerned (cf. Article 27 of the Planning and Building Act, No. 73/1997). Where a master plan has not been approved for the area, which has been subjected to comment from the Nature Conservation Agency and nature conservation committee (cf. Article 33) their opinion of the project must have been delivered before permit is granted. Furthermore, the provisions of the Act on Research and Exploitation of Subterranean Resources, No. 57/1998, apply to extraction on land or from or under the seabed within the net-laying area.

Extraction plan

According to Article 48 NL before permit is granted for extraction of minerals to Article 47 NL, a plan must be provided by the party holding the right to the planned extraction giving details, for instance, of the

- quantity and type of material,
- processing time, and
- clean-up of the extraction area.

The extraction area cannot stand unused and without being cleaned up for more than three years. The Nature Conservation Agency can, however, grant exemptions from this provision, provided there are special grounds for a temporary stoppage. Disposal of waste from the extraction area must be as proposed in the plan submitted (cf. Article 48 NCL, and Acts and Regulations on health and hygiene procedures and pollution prevention).

Rehabilitation of extraction areas (Article 49 NL)

When extraction commences the vegetation and surface layers of the earth in the extraction area have to be dealt with in such manner that they can readily be spread over the extraction area again. Once the processing period is complete, the extraction area shall be neatly re-landscaped so as to fit into its surroundings as well as possible. If the rehabilitation is not carried out as provided for in the extraction plan (cf. Article 48 NL), the Nature Conservation Agency can instruct the party holding right to extraction to carry out the clean-up within a specified time limit, which cannot, however, be longer than one year. The Agency can apply daily fines to enforce this (cf. Article 73 NL). Should actions taken by the Nature Conservation Agency fail to achieve their purpose the local authority has to carry out the clean-up of the extraction area at the cost of the holder of extraction rights in accordance with the plan submitted (cf. Article 48 NL). The guarantee provided for in Article 48 NCL must be used to pay the cost.

17.3.1.3 Authorities

The Nature Conservation Agency has to maintain surveillance of extraction on land (cf. also sub-paragraph b of Article 6 and the second and third paragraphs of Article 7 NL). The Agency can demand that the party holding the right to extraction provide a guarantee which the Agency considers sufficient to cover the estimated cost of surveillance and clean-up of the extraction area.

17.3.1.4 Fees and Taxes

17.3.2 Additional Legal Basics

Table 66: List of Acts significant for raw materials – Iceland

Name of Law	No. of Law and Year of Issuing
Emissions of Greenhouse Gases Act	No. 65 of 2007
Planning and Building Act	No. 73 of 1997
Act on Research and Exploitation of Subterranean Resources	No. 57 of 1998
The Public Procurement Act	No. 84 of 2007
Act on the Right of Ownership and Use of Real Property	No. 19 of 1966 last amended 2002
Act on European Companies	No. 26 of 2004
Act respecting Foundations Engaging in Business Operation	No. 33 of 1999
Competition Law	No. 44 of 2005
The Industrial Act	No. 42 of 1978
Act on Working Environment, Health and Safety in Workplaces	No. 46 of 1980
Law on commodity tax	No. 97 of 1987
Environmental Impact Assessment Act	No. 106 of 2000
The Nature Conservation Act	No. 44 of 1999
Act on Protective Measures Against Avalanches and Landslides	1997
Planning and Building Act	1999

Emissions of Greenhouse Gases Act

The objective of the Emissions of Greenhouse Gases Act is to create conditions for the government to abide by Iceland's international obligations on the limitation of greenhouse gas emissions

18 Ireland

18.1 General Facts

The national territory covers 70 273 km² and the inhabitants are 4,4 million. The population density is 63/km². The GDP per capita for 2009 is estimated at $ 51.356.

Constitutional Structure

The form of government is a republic. Ireland is divided into the national level, district level (29 districts) und municipal level (85 municipalities).

18.2 Production of Raw Materialss

Table 67: Production Data – Ireland (World Mining Data, Weber and Zsak, 2008)

Resources		2002	2003	2004	2005	2006	Change 02/06	Change 05/06
Lead	(t)	32.000	50.300	63.800	72.200	61.800	93,13	−14,4
Zinc	(t)	276.700	419.000	438.300	445.400	425.800	53,89	−4,40
Silver	(kg)	5.000	8.500	6.500	6.300	4.100	−18,00	−34,92
Gypsum	(t)	500.000	500.000	520.000	500.000	450.000	−10,00	−10,00
Natural Gas	(Mm³)	980	960	950	1.000	1.100	12,24	10,00

Table 68: Aggregates Production – Annual statistics/Ireland, 22 April 2008, quantity in million tonnes (UEPG 2008)

Sand & Gravel (1)		Crushed rock (2)		Marine Aggregates		Recycled Aggregates (3)		Manufactured Aggregates (4)	
2005	2006	2005	2006	2005	2006	2005	2006	2005	2006
54,0	n.a.	79,0	n.a.	n.a.	n.a.	1,0	n.a.	0,0	n.a.

(1) Sand and Gravel: sold production including crushed gravel
(2) Crushed rock: sold production (excluding crushed gravel)
(3) Recycled Aggregates: materials coming from construction and demolition waste used in aggregates market
(4) Manufactured aggregates include blast-furnace-slag, electric-arc-furnace-slag, incinerator bottom ash (IBA), pulverised fuel ash (PFA)

18.3 Normative Basics

18.3.1 Primary Legal Basics

The primary legal basics of mineral extraction activity is the Mining Law, which consists of the Mineral Development Acts 1940 to 1949. These comprise:

- Minerals Development Act 1940 (No 31 of 1940)
- Petroleum and other Minerals Development Act (No 7 of 1960)
- Minerals Development Act 1979 (No 12 of 1979)
- Minerals Development Act 1995 (No 15 of 1995)
- Minerals Development Act 1999 (No 21 of 1999)

These Acts must be read together. In addition the Energy (Miscellaneous Provisions) Act 2006 (no. 40/2006) contains important provisions related to rehabilitations of abandoned mineral extraction sites. A new Minerals Development Act to consolidate and update all these laws is scheduled for publication in 2011[1].

18.3.1.1 General Rules

Table 69: Structure of the Minerals Development Act 1940 (No. 31 of 1940)

Minerals Development Act 1940	
Principal Provisions	Main changes made in later Acts
Part 1	**Preliminary** (Articles 1 to 5)
Defines minerals	Definition of minerals amended by 1979 Act to exclude stone sand gravel and clay
Sets out what minerals are in full State ownership ("State Minerals")	
Defines working minerals	
Defines ancillary rights	
Part 2	**Right of Entering and Prospecting Unworked Minerals and Grant of Prospecting Licences** (Articles 7 to 13)
Allows Minister to prospect for minerals or allow others to do so under a Prospecting Licence on terms decided by Minister	Amended by 1995 Act to allow renewal of Prospecting Licences
Sets out public notice procedures before issuance of a Prospecting Licence	

---

1 http://www.taoiseach.gov.ie/eng/Taoiseach_and_Government/Government_Legislation_Programme/SECTION_B1.html

Allows Licence holders to enter land	
Prospecting Licence terms determined by Minister	
Imposes a strict liability for damage to land	Strict liability extended by 1960 Act to include damage to mineral deposits and water supplies and to nuisance
Part 3	**Compulsory Acquisition of Unworked Minerals and of Mining Facilities** (Articles 14 to 25)
	Provisions for compulsory acquisition of unworked minerals were repealed by the 1979 Act
Permits Minister to acquire ancillary rights and land when these are necessary to work State minerals. Sets out the process for such acquisitions and requires payment of compensation	
Part 4	**State Minerals** (Articles 26 to 32)
Permits Minister to lease State minerals	
Royalties and other terms to be agreed with each Lessee	
Lessees have rights to enter land to do anything necessary to work the minerals	
Imposes a strict liability for damage to land	Strict liability extended by 1960 Act to include damage to mineral deposits and water supplies and to nuisance
Part 5	**The Mining Board** (Articles 33 to 37)
Sets up Mining Board to adjudicate on compensation payable under the Act	
Part 6	**Facilities for Development of Privately Owned Minerals and Grant of Restrictions on Working Minerals Required for the Support of Land or Buildings** (Articles 38 to 56)
	Sections relating to development of Private Minerals repealed by 1979 Act
Sets out procedures for acquisition and use of ancillary rights for privately owned minerals	
Part 7	**Compensation** (Articles 57 to 73)
Sets out the procedures for assessment of compensation payable under the Act	Matters relating to compensation for acquisition of private minerals repealed by 1979 Act
Part 8	**Miscellaneous and General** (Articles 74 to 83)
Makes unauthorised working of minerals a criminal offence	

Requires drilling and shaft sinking below 20 feet (6m.) to Minister to be reported to the Minister	
Requires Minister to make a (publicly available) six-monthly report to Parliament	
Schedule	
Includes a non-exclusive list of materials included in the definition of minerals	

Classification of minerals[2]

There are two basic categories of earth resources in Ireland: "minerals" as defined in the Mining Law, and other materials. The former are defined as: all substances in, on, or under land) but excluding peat, stone, gravel, sand and clay. The Minerals Development Act contains an indicative schedule of minerals but this is not exhaustive.

Some of the most important minerals comprehended by the Mining Law are: coal, lignite, dolomite and dolomitic limestone, china and ball clay, silica sand, gypsum and anhydrite, barites, fluorspar, tin, lead zinc and copper, silver and gold.

Materials used in the production of aggregates and cement, such as common clay and limestone are not normally regarded as minerals under mining law, but this is subject to interpretation by the courts in specific circumstances and there is a considerable body of case law on the meaning of terms such as stone and clay.

Ownership of mineral rights[3]

Not all minerals are state-owned. An estimated percentage of 35–40% of these scheduled minerals is privately owned. However, the exclusive right to work minerals is vested in the Minister for Communications Marine and Natural Resources on behalf of the state other than minerals which were already being privately worked on the 15th December 1978. Compensation must be paid by the minister to the private minerals owner if they are worked.

As in many other jurisdictions, a distinction must be made between minerals as defined in Irish law, and other earth resources. The latter comprise stone, sand and gravel, clay and peat, and the ownership of these is a matter for

---

2 Ike, P. (2004), Country Report Ireland, in: Department of Mining and Tunnelling, lc.
3 Ibidem.

ordinary property law. Their ownership is often, but not universally, linked to the ownership of the land and is usually in private hands. Many but not all minerals in the strict sense of the term are in State ownership, including all mines of gold and silver. The right to work minerals is controlled by the Mineral Development Acts, and, with a very few site specific exceptions, is vested in the State, whether or not they are State-owned.

To carry out exploration activities for all minerals comprehended by the Mining Law, it is necessary to obtain a prospecting licence from the State. In addition to a prospecting licence a separate mining lease or licence is also required to extract these minerals, including an agreement on royalties payable to the state.[4]

18.3.1.2 Issuing of Permits

Prospecting und Exploration

Under the Mining Law it is necessary to obtain a prospecting permit prior to carrying out exploration activities. In Ireland Prospecting Permit Competitions are held every three months for those permits that have been surrendered, terminated, or offered but declined in the previous three months. Parties who are interested have 2 months to apply for these competition areas.[5] The list with prospecting permits is published on the first of: February, May, August, and November.

Publications made on the first of May and November also include updating of current licence details, industry news, revised map of State Mining and Prospecting Areas etc. The publications made on the first of February and August are not that extensive. They provide a competition list of prospecting licenses issued or offered and advertised since the previous publication.[6]

All other areas can be applied for at any time and are issued on a first-come first-served basis.

No permit is required for the exploration of other materials but it may be necessary to obtain Planning Permission under the Planning and Development Acts whereas mineral exploration is exempt from this requirement.

---

4   The category "materials" is regulated in environmental law.
5   Ike, P. (2004), Country Report, Irland, in: Department of Mining and Tunnelling, University of Leoben.
6   Ibidem.

Permit applications can take place on-line. Applicants are required to fulfil the following data:

1. proof of technical and financial resources;
2. exploration reason;
3. suitable work programme;
4. request for minimum land use,
5. financial reserves

Exploitation

A State Mining Lease (for State Minerals) or a State Mining Permit (for privately-owned minerals) must be obtained from the Minister for Communications Energy and Natural Resources before minerals are worked. These regulate ownership issues and are individually negotiated. Applications are only accepted from the holder of a valid Prospecting Permit over the relevant area (or State Mining Lease or Permit for an existing mine). Common conditions include:

- Period of validity, which is normally related to the projected mine life
- Royalties and Permit fees varying from about € 0.5 to € 1 tonne for industrial minerals, to a percentage of revenue for metal mines (current examples include 1.75 % for private minerals and 3.5 % for State Minerals).
- Recoupment to the State of any compensation payable to private mineral owners in recognition of which a lower royalty is charged.
- Best practice to be used and extraction of the mineral deposits to be maximised as far as is technically and economically practicable.
- Sureties and funds to be available for closure and rehabilitation
- Leases or permit revocation termination and surrender procedures
- Proper record keeping an provision of information to the Minister

The extraction permit is obtained from the competent authority under the Planning and Development Acts. This is initially the relevant Local Authority, i.e. district level, but there is a right of appeal to a national planning appeals board (An Bord Pleanala). One of the most important requirements is an Environmental Impact Statement which must be prepared by the developer, containing an analysis of the likely effects of the project on people, flora, fauna, soil, water, landscape etc. This must be prepared for all minerals subject to the Minerals Development Acts and for other proposed surface extraction involving more than 25 ha. Concerning conditions, the authorities have broad freedom. These conditions though have to be reasonably related to the development. Common planning conditions are:

- replacement of water supplies affected by the development
- upgrading roads
- landscaping
- control and monitoring of subsidence
- provisions for closing the operation once the deposit has been mined out

The validity of mining permits varies. If unused, the permit is valid for a period of 5 years. It will normally last for the lifetime of the development if a mineral development commences within the 5 year period following the granting of extraction permit. Environmental aspects for all minerals is controlled by an Integrated Pollution Prevention and Control Licence issued by the Environmental Protection Agency under the Environmental Protection Agency Act, 1992 as amended by the Protection of the Environment Act, 2003. Such licences address water and air pollution, and waste management. For other materials these are controlled by licences issued by Local Authorities under the Air Pollution Act 1987, the Water Pollution Acts 1977 and 1990 and the Waste Management Acts 1996 to 2005.

Rehabilitation

The Environmental Protection Agency published a document called "*Environmental Management in the Extractive Industry; Environmental Management Guidelines*" (2003). This document refers to a number of publications providing guidelines for the restoration of quarry developments. Guidelines on restoration and aftercare listed are[7]:

- Consider and develop a restoration scheme at the earliest possible stage in the planning of quarry developments;
- Consult with interested parties regarding after use/restoration options;
- Implement progressive restoration, where possible;
- Maximise soil recovery during stripping operations, and store topsoil and overburden materials separately;
- Retain topsoil and overburden to ensure the materials can be re-used in restoration;
- Provide an appropriate programme of maintenance and aftercare

Closure plans

Major mineral developments will be required to produce detailed closure plans for approval, and provide effective sureties to guarantee that sufficient funds will be available at all times for rehabilitation and aftercare. These plans will be periodically reviewed during the lifetime of the operation, including

---

7　Ibidem.

revision of the surety where necessary. The responsibility for restoration and aftercare lies with the extractive industry, unless assumed by a Public Authority. The Energy (Miscellaneous Provisions) Act empowers the Minister for Communications, Energy and Natural Resources to rehabilitate abandoned mine sites.

18.3.1.3 Authorities

The main responsible authority for mining is the Minister for Communications, Marine and Natural Resources. The Exploration and Mining Division of the Department of Communication Marine and Natural Resources is charged with the application of the Minerals Development Act to minerals exploration and development; the encouragement of the early identification and responsible development by private investors of the Nation's minerals deposits in accordance with best international practice; enhancing the attractiveness of Ireland for international and national minerals investment by active promotional measures.[8]

Mineral operations are periodically inspected by the Exploration and Mining Division and the Health and Safety Authority as well as by local authorities and the Environmental Protection Agency.

The agency responsible for the enforcement of Integrated and Pollution Prevention and Control Licences or permit conditions is the Environmental Protection Agency. It has the duty to supervise the statutory environmental monitoring carried out by local and other prescribed public authorities. The Environmental Protection Agency (EPA) prepares and implements its own environmental monitoring programmes. The EPA is bound to:

- Provide public access to all its monitoring results.
- Establish and maintain a database of information on environmental quality to which the public will have access.

Publish regular reports on the state of the environment.

Local Authorities have extensive powers to enforce the terms of planning permits, and to take action against any unauthorised developments.

Exploration programmes under prospecting permits are approved by the Department if Communications Marine and Natural resources, and failure to follow these may lead to forfeiture of the permit. Mining Leases and Permits contain provisions by which the Minister can require work to be done. Breaches of their terms can lead to their termination.

---

8 http://www.dcenr.gov.ie/Natural/

18.3.1.4 Fees and Taxes

If a mining licence is required under the Mining Law, an application fee is charged. This varies from € 6.300 for small industrial minerals operations to € 19.000 plus € 0.13per tonne of planned annual output for metalliferous mines.[9]

Profits from the extractive industry are subject to a corporation tax rate of 25 %, but some, principally metallic minerals attract favourable capital allowances such as immediate write-off of exploration and development expenditures. Costs of mine rehabilitation after closure can be claimed against profits made before mine closure.

18.3.2 Additional Legal Basics

Table 70: List of Acts significant for raw materials – Ireland

Name of Law	No. of Law and Year of Issuing
Air Pollution Act	Law No. 6 of 1987
Protection of the Environment Act	Law No. 27 of 2003
Waste Management Act	Law No. 10 of 1996
Waste Management (Amendment Act)	Law No 36 of 2001
Planning and Development Regulations	SI 600 of 2001
Planning and Development Act	Law No 30 of 2000
Planning and Development (Amendment) Act,	Law No 32 of 2002
Taxes Consolidation Act	1997
Local Government (Water Pollution) Acts	Law No.1 of 1977 and No. 21 of 1990
Environmental Protection Agency Act 1	Law No. 7 of 1992
Irish Land Act	
Land Act	Law No 42 of 1923
Foreshore Act	Law No 12 of 1933
Safety, Health and Welfare at Work Act	Law No 10 of 2005
Safety, Health and Welfare at Work Act (Quarries) Regulations	SI 28 of 2008
Minerals Development Regulations	SI 340 of 1979
Energy (Miscellaneous Provisions) Act	Law No 40 of 2006
Waste Management (Management of Waste from the Extractive Industries) Regulations	SI 566 of 2009

---

9   Ike, P. (2004), lc.

19 Italy

19.1 General Facts

The national territory covers 301 338 km² and the inhabitants are 60.157.214 (2009 estimation). The population density is 199,6/km². The GDP per capita for 2009 is estimated at $ 35.435.

Constitutional Structure

The form of government is a parliamentary republic. It is divided into 20 regions, of which 5 have a special autonomous status. It is further divided into 109 provinces and 8.100 municipalities.

19.2 Production of Raw Materials

Table 71: Production Data – Italy (World Mining Data, Weber and Zsak, 2008)

Resources		2002	2003	2004	2005	2006	Change 02/06	Change 05/06
Manganese	(t)	960	950	930	920	910	−5,21	−1,09
Aluminium	(t)	190.400	191.400	195.400	192.900	194.200	2,00	0,67
Cadmium	(t)	391	22	0	0	0	−1,0.00	
Lead	(t)	3.200	2.700	800	800	800	−75,00	0,00
Gold	(kg)	631	0	0	0	0	−1,0.00	
Silver	(kg)	0	200	100	100	100		0,00
Baryte	(t)	15.000	15.000	15.500	15.800	10.000	−33,33	−36,71
Bentonite	(t)	250.000	280.000	400.000	500.000	470.000	88,00	−6,00
Feldspar	(t)	1.800.000	2.000.000	2.500.000	2.500.000	3.000.000	66,67	20,00
Fluorspar	(t)	60.000	62.000	64.000	66.000	50.000	−16,67	−24,24
Gypsum	(t)	1.350.000	1.400.000	1.200.000	1.200.000	1.200.000	−11,11	0,00
Salt	(t)	3.600.000	3.800.000	3.600.000	3.600.000	3.700.000	2,78	2,78
Sulphur	(t)	300.000	390.000	570.000	650.000	680.000	1,6.67	4,62
Talc	(t)	100.000	130.000	150.000	180.000	140.000	40,00	−22,22
Steam-Coal	(t)	163.000	250.000	98.000	90.000	70.000	−57,06	−22,22
Lignite	(t)	250.000	240.000	200.000	200.000	0	−1,0.00	−1,0.00
Nat. Gas	(Mm³)	15.100	13.500	12.500	12.000	10.000	−33,77	−16,67

Resources		2002	2003	2004	2005	2006	Change 02/06	Change 05/06
Oil	(t)	5.200.000	6.000.000	6.600.000	6.100.000	5.700.000	9,62	−6,56
Oil-Shale	(t)	23.500	23.200	23.000	24.000	27.000	14,89	12,50

Table 72: Aggregates Production – Annual Statistics/Italy, 22 April 2008, quantities in million tonnes (UEPG 2008)

Sand & Gravel (1)		Crushed rock (2)		Marine Aggregates		Recycled Aggregates (3)		Manufactured Aggregates (4)	
2005	2006	2005	2006	2005	2006	2005	2006	2005	2006
225,0	210,0	145,0	135,0	n. a.	0,0	4,5	5,5	3,0	3,5

(1) Sand and Gravel: sold production including crushed gravel
(2) Crushed rock: sold production (excluding crushed gravel)
(3) Recycled Aggregates: materials coming from construction and demolition waste used in aggregates market
(4) Manufactured aggregates include blast-furnace-slag, electric-arc-furnace-slag, incinerator bottom ash (IBA), pulverised fuel ash (PFA)

19.3 Normative Basics

19.3.1 Primary Legal Basics

The primary legal basics of mineral extraction activity is the Mining Law (Regio Decreto) No. 1443 of 1927 as amended by Legislative Decret No. 213 of 1999.

19.3.1.1 General Rules

Table 73: Structure of the Mining Law (Regio decreto, 29 luglio 1927, n. 1443)

Preamble	Articles 1–33
A) Expiration of the term	Articles 34–37
B) Surrender	Articles 38–39
C) Revocation	Articles 40–41
Provisions customary to Surrender and Revocation	Articles 42–65

The exploration and exploitation of minerals are governed by Article 1 ML. Workings provided in Article 1 can be divided into two categories: mining and quarrying. The first one relates to (Article 2):

a) energy minerals;
b) metal-bearing minerals (i. e. tin, lead, zinc, copper, silver, gold);
c) non-metal-bearing minerals of significance industrial importance (i. e. salt and potash, Fuller's earth, baryte, fluorspar).

The second category includes construction minerals and other industrial materials not included in the first category.

Ownership of mineral rights

The rights of ownership and exploitation of a mineral resource depends on the category under which it is listed[1]:

- The State owns the rights to first category minerals and, through the Ministry of Industry, issues permits for exploration, research and exploitation. Rights to marine sand and gravel also belong to the State.
- Second category minerals are the property of the land owner, and are subject to regional administrative regulations.

However, by a presidential decree, the substances comprised in the second category can be included in the first category.

19.3.1.2 Issuing of Permits

Exploration

Exploration is allowed only for those who hold the permit (Article 4). Permit to explore cannot be arranged for more than 3 years. However, this may be extended upon confirmation, at the expense of the explorer (Article 6).

The explorer must pay the State a prescribed tax for each used hectare of land within the limits of the permit (Article 7).

The exploration permit is not transferable without the prior consent of the Minister of Industry, Trade and Crafts. Each transfer is subject to a fixed fee (Article 8).

The Minister for Industry, Trade and Crafts can pronounce the forfeiture of the exploration permit (1): when the work has remained suspended for over 3 months or the operator do not comply with the requirements or will be contrary to the provisions of this law (Article 9).

---

1 Badino, G. et al (2004), Country Report Italy, in: Department of Mining and Tunnelling, lc.

Within the limits of the land comprised in an exploration permit (or concession) another exploration permit can be issued, but for different substances (i. e. minerals), and provided new operations are not incompatible with those of existing exploration or concession activities. In case of disagreement between the parties concerned, the Minister of Industry, Trade and craft, after consulting the Board of Mines must make a decision (Article 11).

When the State intends directly to explore, the area of exploration is determined by decree of the Minister of Industry, Trade and Crafts (Article 13).

Exploitation

In no case extraction of minerals can be done without the authorization of the Minister of Industry, Trade and Crafts, which has the competent authority of the mining district (Article 12).

According to Article 14 ML the mines can be exploited only by those who have the grant. Also more concessions in the same area are possible but for different minerals based on requirements by Article 11.

According to Article 15 ML the granting of a concession can be made (by the Minister of Industry, Trade and Crafts) to those which have the technical and economic ability to lead a company (1). More concessions to the same person are possible. For mines laid in areas of interest to the defence, the Minister of Industry, Trade and Crafts provides a concession agreement in consultation with the military administration (2).

The explorer (of a deposit) is preferred to any other applicant, provided that the Minister for Industry, Trade and Crafts finds that he possesses the technical and economic suitability. If the explorer/operator receives the concession, he has the right to obtain a premium (in relation to the importance of the discovery) and compensation on account of the operating system. The premium and the benefits are provisionally established in the act of concession (Article 16).

The concession is issued by decree of the Minister of Industry, Trade and Crafts. The decree contains (Article 18):

a) an indication of the operator and his home or regarding the province where the mine is placed,
b) duration of the concession,
c) nature of the situation, extension of the mine and its boundary,
d) particulars of the right proportion to be paid by the concessionaire under the terms of Article 25

e) amount of reward and compensation to the explorer according Article 16
f) State's participation in the profits of the company, determined after hearing the Minister for Finance,
g) plan of delimitation of the concession.

When the award has been made to the operator, he must within 3 months from the date of communication of the decree granting, contact the Ministry of Industry, Trade and Crafts, by submitting a certificate of having paid the explorer the amount specified in the decree. Failure to comply with these requirements, result in the forfeiture of the award, pronounced by the Minister of Industry, Trade and Crafts (Article 20 ML).

The granting of the concession is temporary (Article 21). The concession and its appliances are subject to the provisions of law governing the property (Article 22 ML).

According to Article 26 ML the mines must be kept in operation except that the Minister of Industry, Commerce and Crafts, has permitted the suspension of work (1). The option to allow the suspension of work relates to the same authority which issues the extraction permits. That option is always the responsibility of the mining district authority in the case of suspension for periods not exceeding one year (2). The operator must recultivate the mine with technical and economic resources appropriate to the importance of the deposit (3).

Any transfer of the concession must be previously authorized by the Minister of Industry, Trade and Crafts (Article 27 ML). Any act that has not the prior authorization is invalid. For mines lying in areas of interest of national defence, the Minister of Industry, Trade and Crafts authorizes such transfers after hearing the military administration. The decree authorizing the transfer is recorded with a fixed tax.

The mining concessionaires must provide public administration with the statistics and any other information that is asked of them. They must also make available to officials delegated all necessary means to inspect the work (Article 29 ML).

According to Article 31 ML the permit holder is required to check any damage arising from the mine as regards the provision of any security by complying with standards set out in Article 10 (1) ML.

Within the perimeter of the concession, works needed for the storage, transport and processing of materials for the production and transmission, and generally for the safety of the mine, are considered public utilities. In case of dispute about the need for modalities of the works mentioned above, the decision is taken by the competent authority (chief engineer of the mining district) (Article 32 ML).

Expiration of the validity

The concession shall cease: a) on expiry of the period, b) for waiver or c) for revocation (Article 33). The concession expired can be renewed if the permit holder has complied with their obligations (Article 34). If the concession is not renewed, the permit holder has, upon expiry of the term, close the mine (Article 35).

Quarries

The quarries are left to the ground owner (Article 45). When the operator fails to undertake the extraction, the competent authority (chief engineer of the mining district) can set a deadline for beginning the recovery or the intensification of work. If the deadline is not observed, the chief engineer of the mining district can issue the extraction permit of the quarry in accordance with the rules in this law, as applicable.

When mining/quarrying cause any damage a compensation between the concerned parties is required (Article 46).

Recultivation of mines and quarries

Implementation, maintenance and use of any work needed for a useful recultivation of mines and quarries can be voluntary or mandatory consortia. The consortium is required by Decree of the Minister of Industry, Commerce and Crafts (i. e. Board of Mines) (Article 47).

Mandatory and voluntary consortia of mines and quarries will be granted by Decree of the Minister of Industry, Trade and Crafts and will be collected as tax privileges. All acts that are engaged directly in the interests of these consortia are registered with a fixed fee. The income from the mines/quarries or works carried out by consortia will be for 20 years from the date of completion of this work and will be exempt from property tax (Article 48). If within the time limits, the works are not performed, the Minister of Industry, Trade and Crafts has to appoint a commissioner who, at the expense of the operator assumes the administration (Article 49).

19.3.1.3 Authorities

The main responsible authority for mining is the Minister of Industry, Trade and Crafts. With specific regard to mineral extraction, the Ministry of Industry is responsible for controlling mineral activities. Its functions are performed by the Ministry's Regional Offices, which are responsible for granting permits.[2]

The regions were given responsibility for overseeing extraction activities for the second category minerals under DPR 14 January 1972, n. 2, and DPR 24 July 1977, n. 616. After, each region was required to issue its own laws to regulate and manage second category mineral production. This has led to widely differing approaches being adopted by the different regions.

19.3.1.4 Fees and Taxes

The authorization according to Article 12 ML is subject to government license. The operator is required to pay annually to the State a tax for each used hectare of land within the limits of the concession (Article 25).

The regional or provincial authority will levy a tax on the operator, which is paid to the municipality, based on the quality and quantity of material extracted each year. The proceeds from this tax will be released as necessary for maintenance works.[3]

Regional legislation specifies that operators must enter an agreement with the communal authorities, requiring them to carry out restoration works at the operator's expense, and that provisions can be made for ownership to be transferred over to the commune once activity has ceased, if this is the end-use agreed in the excavation plan. This agreement must also provide for an obligation on the part of the applicant to pay the communal authority an annual sum known as "excavation fees" commensurate with the quantities and type of material extracted, according to tariffs established by the Regional Council. 80 % of the receipts go to the municipality (to help pay for monitoring and enforcement of restoration), 15 % to the provinces (to help pay for minerals research and administration) and 5 % to the region. In practice a large proportion of the fees are set aside for maintenance following restoration for which the applicant is responsible. The size of the fees depends on the mineral. However, the first

---

2   Badino, G. et al (2004), Country Report Italy, in: Department of Mining and Tunnelling, lc.
3   Ibidem.

category minerals (all minerals excluding i.e. construction minerals) of this law are not subject to this tax.[4]

19.3.2 Additional Legal Basics

Table 74: List of Acts significant for raw materials – Italy (Badino, G. et al, 2004)
National Legislation

	No. of Law and Year of Issuing	Changes
Framework Law for Protected Areas	Law No. 394 of 1991	
Regulations for the Improvement of Soil Protection	Law No. 183 of 1989	Replaced by Legislative Decree 3 No. 152 of l 2006
Regulations concerning Environmental Damage	Law No. 349 of 1986	
Provisions for the Conservation of Areas of special Environmental Interest	Law No. 431 of 1985	Deleted. Today in force Legislative Decree No. 42 of 2004, "Code of the cultural heritage and landscape, in accordance with article 10 of L.R 6 July 2002, n. 137"
Regulations for the Protection of Water	Law No. 319 of 1976	
Decree: Transfer of the National Administrative Functions relating to Minerals, Quarries to the Regions	DPR 14 January 1972, n. 2	
Decree: Ministry of the Productive Activities. Approval of the Directory of the recognized Products suitable in Developing to the Employment in the extractive Activity. (GU n. 40 of the 18-2-2004 - Suppl. Ordinary n.25)	Decree 2004	
Decree: Ministry of the Productive Activities. Integration of the Directory of the interesting Areas for the Operating Mining Research. (GU n. 203 of 2-9-2003)	Decree 2003	

---

4   Ibidem.

19.3 Normative Basics

	No. of Law and Year of Issuing	Changes
Decree: Modalities and Criteria of Access to the Facilities for the Restructure and the Structural Modification of Atmospheres of Jobs in the Quarry localized in metamorphic Limestone Deposit with Development to Quotas beyond 300 meters of which all Article 114, codicil 4, of the Law 23 December 2000, n. 388. (the 18 G.U. n. 217 of September 2001)	Decree 2001	
Legislative Decree: To act RD (Regio Decreto) 29 July 1927, n. 1443	Decree 1999 No. 213	
Law: New Norms for the Performance of Mining Politics	Law No 221 of 1990	
DPR: Implementation of Delegation as article 1 of Law 1975. (which, when taken together, transferred Responsibility for Regulating 2nd category Mineral Activities to the Regions)	DPR No. 6 of 1977	
D.P.R. 18-4-1994 n. 382 Disciplina dei procedimenti di conferimento dei permessi di ricerca e di concessioni di coltivazione di giacimenti minerari di interesse nazionale e di interesse locale. Pubblicato nella Gazz. Uff. 18 giugno 1994, n. 141	DPR No. 382 of 1994	
Decreto Legislativo 9 aprile 2008 , n. 81 Attuazione dell'articolo 1 della Lr 3 agosto 2007, n. 123, in materia di tutela della salute e della sicurezza nei luoghi di lavoro	Decree No. 81 of 2008	
Decreto Legislativo 25 novembre 1996, n. 624 "Attuazione della direttiva 92/91/CEE relativa alla sicurezza e salute dei lavoratori nelle industrie estrattive per trivellazione e della direttiva 92/104/CEE relativa alla sicurezza e salute dei lavoratori nelle industrie estrattive a cielo aperto o sotterranee"– Pubblicato nella Gazzetta Ufficiale n. 293 del 14 dicembre 1996 – Supplemento Ordinario n. 219	Decree No. 624 of 1996	
D.Lgs. n. 117/2008,Attuazione della direttiva 2006/21/CE relativa alla gestione dei rifiuti delle industrie estrattive e che modifica la direttiva 2004/35/CE	No. 117 of 2008	
Decreto 30 giugno 2009: Ministero dello Sviluppo Economico. Disciplina e modalita' di attribuzione di giacimenti concessionari di coltivazioni di idrocarburi marginali. (GU n. 169 del 23-7-2009)	Decree No. 169 of 2009	

	No. of Law and Year of Issuing	Changes
Decreto 16 febbraio 2006: Ministero delle Attività Produttive. Rinnovo della commissione interdisciplinare consultiva per la ricerca mineraria di base. (GU n. 55 del 7-3-2006)	Decree No. 55 of 2006	
Decreto 3 novembre 2005: Ministero delle Attività Produttive. Criteri per la determinazione di un adeguato corrispettivo per la remunerazione dei beni destinati ad un concessionario per lo stoccaggio di gas naturale, ai sensi dell'articolo 13, comma 9, del decreto legislativo 23 maggio 2000, n. 164. (GU n. 272 del 22-11-2005)	Decree No. 272 of 2005	
Decreto 13 luglio 2005: Ministero delle Attività Produttive. Integrazione all'elenco delle aree indiziate per la ricerca mineraria operativa, ai sensi degli articoli 5 e 6 della Lr 6 ottobre 1982, n. 752. (GU n. 172 del 26-7-2005)	Decree No. 172 of 2005	
Decreto 10 giugno 2005: Ministero delle Attività Produttive. Modifiche all'elenco dei prodotti esplodenti riconosciuti idonei all'impiego nelle attività estrattive. (GU n. 145 del 24-6-2005)	Decree No. 145 of 2005	
Decreto del Presidente del Consiglio dei Ministri 9 giugno 2005: Trasferimento delle risorse finanziarie e umane per l'esercizio delle funzioni in materia di miniere e risorse geotermiche alla regione Sardegna. (GU n. 154 del 5-7-2005)	Decree No. 154 of 2005	
Decreto 16 novembre 2004: Ministero delle Attività Produttive. Modifiche all'elenco dei prodotti esplodenti riconosciuti idonei all'impiego nelle attività estrattive. (GU n. 288 del 9-12-2004)	Decree No. 288 of 2004	
Decreto 21 Gennaio 2004: Ministero delle Attività Produttive. Approvazione dell'elenco dei prodotti esplodenti riconosciuti idonei all'impiego nelle attivita' estrattive. (GU n. 40 del 18-2-2004 - Suppl. Ordinario n.25)	Decree No. 40 of 2004	
Decreto 7 agosto 2003: Ministero delle Attività Produttive. Integrazione all'elenco delle aree indiziate per la ricerca mineraria operativa ai sensi degli articoli 5 e 6 della Lr 6 ottobre 1982, n. 752. (GU n. 203 del 2-9-2003)	Decree No. 203 of 2003	

	No. of Law and Year of Issuing	Changes
Decreto 8 giugno 2001: Modalità e criteri di accesso alle agevolazioni per la ristrutturazione e la modifica strutturale degli ambienti di lavoro nelle cave localizzate in giacimenti di calcare metamorfico con sviluppo a quote di oltre 300 metri di cui all'Article 114, comma 4, della Lr 23 dicembre 2000, n. 388. (G.U. n. 217 del 18 settembre 2001)	Decree No. 217 of 2001	
RL 30 luglio 1990, n. 221: Gazz. Uff., 7 agosto 1990, n. 183 Nuove norme per l'attuazione della politica mineraria	No. 221 of 1990	
D.P.R. 9-4-1959 n. 128 Norme di polizia delle miniere e delle cave	DPR No. 128 of 1959	
Regio decreto 29 luglio 1927, n. 1443: (in Gazz. Uff., 23 agosto 1927, n. 194). Norme di carattere legislativo per disciplinare la ricerca e la coltivazione delle miniere [nel Regno]. Il presente decreto è aggiornato e coordinato al d.lgs. 4 agosto1999, n.213	Royal Decree No. 1443 of 1927, amended by No. 213 of 1999	

Table 74a: Regional laws on quarrying activities (RL means "Regional Law"; PL means "Provincial Law")[5]

Region or Autonomous Province	Law Number and Date	
Piemonte	RL 69/1978 – RL 44/2000	RL n 30 del 03/12/1999 RL n. 28 del 30/04/1996 RL n. 30 del 12/08/1981 RL n. 9 del 13/03/1981 RL n. 6 del 18/02/1980 RL n. 69 del 22/11/1978
Valle d'Aosta	RL 44/1982 – RL 15/1996	
Lombardia	RL 14/1998	
Provincia Autonoma di Trento	PL. n. 7, 24 October 2006	
Provincia Autonoma di Bolzano	Lp. 32/1976 – Lp. 7/2003	
Veneto	RL 44/1982 – RL 15/1983	
Friuli Venezia Giulia	RL 35/1986 – RL 25/1992	
Emilia Romagna	RL 17/1991	RL n. 21 aprile 1999, n. 3 RL n. 7 del 14-04-2004 RL n.17 del 18 luglio 1991

5   Information provided by Ubaldo Cibin (Geological Survey[Bologna/Italy]).

Region or Autonomous Province	Law Number and Date	
Liguria	RL 12/1979 – RL 21/2001	RL 27.09.2002, n. 34 RL 24 luglio 2001 n. 21 RL 01/09/1995 n. 46 RL 30/12/1993 n. 63 RL 10/04/1979 n. 12.
Toscana	RL 78/1998	Decreto del presidente della giunta regionale del 23-02-2007 n. 10; RL n. 4 del 27-01-2004 RL del 03/11/1998 n. 78
Umbria	RL 2/2000 – RL 34/2004	RL n. 34 del 23-12-2004 RL del 15/01/2001n. 3 RL 3 gennaio 2000, n. 2
Marche	RL 33/1999 – RL 15/2003	RL n. 15 del 30-06-2003 RL 24 luglio 2002 n. 14 RL 17/12/1999 n. 33 RL 1 dicembre 1997, n. 71
Lazio	RL 27/1993 - RL 17/2004	RL n. 17 del 6 dicembre 2004 RL del 30 novembre 2001, n. 30 RL del 05/05/1993 n. 27
Abruzzo	RL 54/1983 – RL 8/1995	RL del 26/07/1997 n. 69 RL del 23/10/1987 n. 67 RL del 26/07/1983 n. 54
Molise	RL 11/2005	
Campania	RL 54/1985 – RL17/1995	RL 13 aprile 1995, n.17 RL del 13/12/1985 n. 54
Puglia	RL 37/1985 – RL 21/2004	
Basilicata	RL 12/1979 (updated by Lr 17/2005).	RL n. 21 del 1.3.2005 RL n. 19 del 25.2.2005 RL del 31/03/1980 n. 18 RL del 27/03/1979 n. 12
Calabria	None	
Sicilia	RL 127/1980 – RL 19/1995	RL n. 10 del 5-07-2004 RL 16 aprile 2003, n. 4 RL 3 luglio 2000, n. 14 RL del 06/10/1999 n. 25 RL 6 ottobre 1999, n. 25 RL del 01/03/1995 n. 19
Sardegna	RL 30/1989 – RL 28/1991	RL del 02/06/1994 n. 26 RL n. 28 del 08/08/1991 RL del 11/06/1990 n. 16

20 Kosovo

20.1 General Facts

The national territory covers 10.887 km^3 and the inhabitants are about 2,1 million. The population density is 190/km^2. The GDP per capita is estimated at $ 2.965 (2009).

The Kosovo was administered by the United Nations Organization from 1999 to 2008. In 2008, Kosovo declared independence. Until present, the Kosovo has been recognized as an independent state by 70 states, including 22 member states of the European Union.

20.2 Production of Raw Materials

Table 75: Production data – Kosovo. (Metric tons unless otherwise specified), (USGS, 2008)

Resources		2004	2005	2006
Ferroalloys, ferronickel (38% Ni), gross weight		–	–	–
Lead-zinc:				
Ore, gross weight		–	12.200	63.517
Lead content of ore		–	800	3.900
Zinc content of ore		–	700	3.800
Nickel:				
Ore, wet		–	–	–
Ni content of FeNi		–	–	–
Clay, bentonite			–	
Marl	m^3	202.094	264.814	254.386
Sand and gravel, excluding glass sand	m^3	15.046	14.894	34.268
Lignite	1000 metric tons	5.658	6.391	6.532

20.3 Normative Basics

20.3.1 Primary Legal Basics

The primary legal basic of mineral extraction activity is the Law No. 03/L-163 on Mines and Minerals. This law replaced UNMIK Regulation No. 2005/2 on the Establishment of the Independent Commission of Mines and Minerals and UNMIK Regulation No. 2005/3 on Mines and Minerals.

20.3.1.1 General Rules

Table 76: Structure of the Law on Mines and Minerals (LMM) – Kosovo

Part I Purpose, Scope and Definitions	Article 1: Purpose	
	Article 2: Scope	Item 1–2
	Article 3: Definitions	Item 1–6
Part II General Provisions	Article 4: Authority Required for Exploration or Mining	Item 1–2
	Article 5: Types of License and Permits	
	Article 6: Eligibility Criteria	
	Article 7: Generally Applicable Procedures	Item 1–7
	Article 8: Surrender of License Area	Item 1–6
	Article 9: Suspension or Cancellation of License or Permit for Non-Compliance	Item 1–5
	Article 10: Transfer and Pledge of Licenses and Permits	Item 1–11
	Article 11: Transfer of Control	Item 1–2
	Article 12: General Commercial Rights of Licensees and Permit Holders	
	Article 13: Restricted Activities	Item 1–2
	Article 14: Exercise of Rights	
	Article 15: Use of Wasteful Practices	Item 1–3
	Article 16: Required Office in Kosovo	
	Article 17: Reports, Records and Information	
	Article 18: Licenses Subject to Competitive Award Process	Item 1–9
	Article 19: Basic Requirements for the Competitive Award Process	
	Article 20: Register of Mining Licenses	

Part III Exploration Licenses	Article 21: Scope and Term	Item 1–5
	Article 22: Applications	
	Article 23: Criteria for Issuance	Item 1–4
	Article 24: Exploration License Rights and Obligations	Item 1–10
	Article 25: Exploration License Extension	Item 1–3
Part IV Retention Licenses	Article 26: Scope and Term	Item 1–4
	Article 27: Application	Item 1–2
	Article 28: Criteria for Issuance	Item 1–2
	Article 29: Rights and Obligations	Item 1–2
Part V Mining Licenses	Article 30: Term and Scope	Item 1–2
	Article 31: Applications	
	Article 32: Application Processing	Item 1–3
	Article 33: Criteria for Issuance	Item 1–2
	Article 34: Mining License Rights and Obligations	Item 1–7
	Article 35: Suspension of Production	Item 1–2
	Article 36: Extension of a Mining License	Item 1–4
Part VI Permits for Special Operations	Article 37: Term and Scope	Item 1–2
	Article 38: Application	
	Article 39: Application Processing	Item 1–2
	Article 40: Criteria for Issuance	Item 1–2
	Article 41: Special Permit Rights and Obligations	Item 1–5
	Article 42: Suspension of Operations	Item 1–2
Part VII Small Scale Artisanal Mining	Article 43: Term and Scope	
	Article 44: Application	Item 1–3
	Article 45: Application Processing	Item 1–2
	Article 46: Criteria for Issuance	
	Article 47: Municipal Rights and Obligations	Item 1–4
	Article 48: Authority to Exclude Users	
Part VIII Public Interest Licenses and Public Interest Permits	Article 49: Issuance of Public Interest License and/or Public Interest Permit to a Publicly Owned Enterprise or Socially Owned Enterprise	Item 1–9
Part IX Royalties and Levies	Article 50: Royalties	Item 1–7
	Article 51: Deferments and Exemptions	Item 1–3
	Article 52: Payment of Estimated Royalty	Item 1–2
	Article 53: Failure to Pay Royalty	Item 1–2
	Article 54: Administrative Fees	

Part X The Independent Commission for Mines and Minerals	Chapter I Independent Commission for Mines and Minerals: Establishment, Functions and Organization	Article 55: The Establishment of the Independent Commission for Mines and Minerals	Item 1–5
		Article 56: Powers and Responsibilities of the ICMM	Item 1–3
		Article 57: Implementing Regulations	
		Article 58: Reporting	
	Chapter II The ICMM Board: Establishment, Functions and Organization	Article 59: ICMM Board	Item 1–10
		Article 60: Removal of Members of the Board	Item 1–4
		Article 61: Conflict of Interest	Item 1–5
		Article 62: Organization and Voting	Item 1–6
		Article 63: Director	Item 1–7
		Article 64: Budgeting and Financing	Item 1–3
		Article 65: Dedicated Revenue	Item 1–2
		Article 66: Confidentiality	
Part XI Mining Inspectorate	Article 67: The Mining Inspectorate		Item 1–2
	Article 68: Chief Inspector		Item 1–6
	Article 69: General Authority of the Mining Inspectors		
	Article 70: Illegal Exploration, Mining and Special Operations		Item 1–4
	Article 71: Authority to Enforce Environmental and Safety Legal Provisions		Item 1–2
	Article 72: Investigative Authority of the Mining Inspectorate		Item 1–4
Part XII The Ministry of Energy and Mining	Chapter I Mining Strategy and Mineral Resource Management Plan	Article 73: Mining Strategy	Item 1–4
		Article 74: Mineral Resources Management	Item 1–2
	Chapter II Mining Safety and Environmental Protection	Article 75: Mining Safety regulations	
		Article 76: Mining Environmental Protection Regulations	
	Chapter III Geological Institute of Kosovo	Article 77: Geological Institute of Kosovo: Establishment and Functions	Item 1–2
	Chapter IV National Museum of Crystals and Minerals	Article 78: National Museum of Crystals and Minerals	Item 1–5

Part XIII Remedial and Penalty Provisions	Article 79: Imposition of Administrative Fines and Other Remedies	Item 1–6
	Article 80: Criminal Penalties for Illegal Mining Activities	Item 1–2
	Article 81: Penalties for Failure to Comply with an Order of the ICMM or a Mining Inspector	Item 1–3
	Article 82: Claims for Damages, Injunctions, and Other Remedies	
Part XIV Transitional and Final Provisions	Article 83: Cooperation between Public Authorities	
	Article 84: Administrative and Judicial Review and Appeal	
	Article 85: Publication of Sub-normative Acts	
	Article 86: Existing Licenses	Item 1–2
	Article 87: Public Money	
	Article 88: Repeal of Prior Legislation	
	Article 89: Entry into Force	

According to Article 3 LMM it has to be distinguished:

"Construction Minerals" means all minerals and associated materials normally exploited for the purposes of and/or use in the construction industry, including sand, gravel, limestone (and other sedimentary rocks), granite (and other magmatic rocks), and marble (and other metamorphic rocks), including clay, marl, gypsum chalk, alum, andesite, basalt, dunite, diabase, , tuff, schist, shale, slate, sandstone, quartz aggregates, and construction rocks serpentine, periodite, dunite, diorite, diabasite, basalt, gabro, syenite, gnejset, and quaritzite.

"Industrial Minerals" includes graphite, corundum, quartz (for industrial purposes), bauxite, olivine, kyanite, andalusite, sillimanite, wollastonite, asbestos, talc, pyrophyllite, muscovite and other micas, vermiculite, kaolin, feldspar, nepheline, leucite, scapolite, apatite, baryte, calcite, dolomite, magnesite, fluorite, sulphur, phosphates, gypsum, bentonite, diatomaceous earth, "fullers" earth, palygorsite, attapulgite, sepiolite, salt, anhydrite, gypsum, perlite, celestite, boron minerals, cryolite and pyrite, and shall include Water as defined hereunder.

"Metallic Minerals" means minerals containing the following metals: lithium, rubidium, caesium, beryllium, magnesium, strontium, radium, boron, aluminium, scandium, yttrium, the rare earth metals (lanthanides), actinium, thorium and other actinides, germanium, tin, lead, arsenic, antimony, bismuth, sulphur, selenium, tellurium, copper, silver, gold, zinc, cadmium, mercury, gallium, indium, thallium, titanium, zirconium, hafnium, vanadium, niobium, tantalum, chromium, molybdenum, tungsten, manganese, rhenium, iron, cobalt, nickel and platinum and metals of the platinum group (PGMs).

Ownership of mineral rights

Article 2 LMM provides that mineral resources, regardless of their origin, shape or physical state that are under or on the surface and within the territory of Republic of Kosovo, are property of the Republic of Kosovo. Mineral resources that are extracted by a mining licensee under a valid mining license that has been issued pursuant to the LMM become the property of the mining licensee as and when such minerals are extracted and deposited on the surface.

The state has the responsibility about issuing mineral rights. Article 6 LMM regulates the criteria a person must fulfill in order to be eligible for obtaining a permit.

The applicant must be a currently registered Business Organisation having at least one senior technical manager responsible for the day-to-day technical operations of the Business Organisation who intends in goodfaith to be physically present in Kosovo.

Article 7 LMM regulates the general principles governing the permission procedure. The Independent Commission on Mines and Minerals (ICMM) issues procedures and standard forms for the application, recordation, processing and issuance of Permits, and for the extension, modification, transfer, suspension or cancellation thereof (Article 7,1 LMM). Article 10 LMM regulates the transfer and pledge of permits. For this purpose the licence holder has to prepare an application.

A permit holder has the following commercial rights and obligations to (Article 12 LMM):

(a) carry on business in accordance with good commercial mining practice and with international safety procedures;
(b) maintain books of account in Euro;
(c) open and operate bank accounts both in Kosovo and abroad;
(d) import directly from his respective suppliers all requisite goods and equipment; and
(e) receive and retain abroad proceeds from export sales.

Furthermore there are areas where mining is not allowed. A permit holder is not allowed to conduct any activity that (Article 13 (1) LMM): (a) is within 200 m of the boundaries of any village, city; (b) or is the site within 100 m of any inhabited, occupied or temporarily uninhabited house or building; (c) is within 45 m of any land that has been ploughed or otherwise prepared for the growing of farm crops.

20.3.1.2 Issuing of Permits

A mining license for energy minerals is valid for no more than thirty (30) years; and is extendable for further terms of up to fifteen (15) years. It applies to such area as required for the concerned mineral resource (Article 30.2 LMM). A mining license for all other minerals is valid for no more than twenty five (25) years; and is extendable for another period of not more than twenty five (25) years. It applies to such area as required for the concerned mineral resource (Article 30 (1) LMM).

Article 31 LMM regulates the application provisions. A person desiring to extract a mineral resource has to submit to the Commission a completed application for the issuance of a mining license for the concerned mineral resources within the concerned area. Such application has to be submitted in the prescribed format, together with the prescribed fee. The following has to be attached thereto Article 31 (1) LMM:

Exploration

Part 3 LMM includes provisions related to exploration licences. The law distinguishes between the categories "construction minerals" and "other minerals".

An exploration license for construction minerals is valid no more than 2 years; and can be extended once only for a maximum of 2 additional years (Article 21 (1) MR). It applies to an area no larger than 250 hectares.

An exploration permit for all other minerals is valid of no more than 3 years; and can be extended a maximum of 3 times, each such extension to be for a period of no more than 2 years; provided, however, that on the occasion of each such extension the area to which the exploration permit then applies must be reduced by at least 50%. The exploration permit applies to an area no larger than 100 km² (Article 21 (2) LMM).

The Commission must establish the reduced area under subsection 2 above after taking into consideration a respective proposal from the applicant. Such proposal must be based on the results of all exploration operations conducted to date (Article 21 (3) LMM).

A person desiring to undertake exploration in Kosovo must submit to the Commission a completed application for the issuance of an exploration permit for the concerned Mineral Resource Group within the concerned area. Such application must be submitted in the prescribed format, together with the prescribed fee (Article 22(1) LMM).

Exploration permit rights and obligations: An exploration permit confers on the permit holder the exclusive right to carry out the specified Exploration Operations (24.1 LMM).

An exploration permit holder must (24.2 LMM): (a) commence Exploration Operations within 3 months from the date of issuance; (b) expend on the specified exploration operations not less than the applicable minimum amount prescribed by the Commission ("Minimum Exploration Expenditure").

Additionally, he must (24 (4) LMM): Keep full and accurate records on: (i) boreholes drilled; (ii) lithologies penetrated, with detailed logs thereof; (iii) mineral resources identified; (iv) the results of all geochemical or geophysical surveys; (v) the results of all analysis or identification of minerals; (vi) the geological interpretation of the records maintained under subparagraphs (i) to (v).

According to Article 24 (4.3) LMM an exploration permit holder must submit to the Commission, within two months from the end of each calendar year: (i) an annual report for such calendar year, describing all exploration operations undertaken, identifying the location of all supporting documents and records, and providing an estimate of all mineral resources identified, illustrated with plans at an appropriate scale; and (ii) annual statements for such calendar year on expenditures, accidents and safety at the work site, and recultivation activities.

An exploration permit holder (4.4) must submit to the Commission, not later than September 30 of each calendar year, an exploration programme for the following calendar year. An exploration permit holder desiring an extension of his permit must submit to the Commission a completed extension application in the prescribed format, together with the prescribed fee (Article 25 LMM).

Exploitation phase

A mining license for energy minerals is valid for no more than thirty (30) years; and is extendable for further terms of up to fifteen (15) years. It applies to such area as required for the concerned mineral resource (Article 30.2 LMM). A mining license for all other minerals is valid for no more than twenty five

(25) years; and is extendable for another period of not more than twenty five (25) years. It applies to such area as required for the concerned mineral resource (Article 30 (1) LMM).

Article 31 LMM regulates the application provisions. A person desiring to extract a mineral resource has to submit to the Commission a completed application for the issuance of a mining license for the concerned mineral resources within the concerned area. Such application has to be submitted in the prescribed format, together with the prescribed fee. The following has to be attached thereto Article 31 (1) LMM:

(a) The names and addresses of the applicant's Significant Owners and directors, if any;
(b) The official registered address of the applicant in Kosovo;
(c) A description of the technical, professional and managerial capabilities of the personnel to be principally responsible for the conduct of the Mining Programme, including a copy of the respective curriculum vitae and all relevant diplomas, permits and certificates;
(d) A map showing:
   (i) The area under application, defined by coordinates;
   (ii) The location of any existing building, infrastructure, agricultural activity or other significant manmade improvement or surface feature within such area or within one kilometre from any of its boundaries;
   (iii) The proposed route of access by the applicant to such area; and
   (iv) The boundaries of each concerned Municipality;
(e) All relevant cadastral documentation as prescribed by the Commission;
(f) the environmental consent issued by the Ministry of Environment and Spatial Planning and any approvals of other public authorities that may be required under Kosovo law;
(g) Surface Rights Agreements, having a duration of not less than the duration of the proposed Mining Permit, between the applicant and all third parties having lawfully registered property rights to the surface area that the applicant proposes to use during the conduct of the concerned Mining Operations, or evidence of refusal or failure of such third parties to enter into such an agreement on reasonable terms;
(h) The mining programme;
(i) A mine closure plan;
(j) A social implication s study;
(k) A performance bond;
(l) A copy of the relevant exploration or retention licence held by the applicant.

Within three (3) months after receiving a complete application for a Mining License from an eligible applicant, the competent authority has either issue the concerned License or provide the applicant with a written explanation why the authority has decided not to issue the concerned License (Article 32 (1) LMM).

According to Article 33 LMM the applicant's proposed mining programme must be based on international practice of mining and environmental protection, and must ensure the efficient and beneficial use of the concerned mineral resources in the concerned area. The applicant must prove the financial resources reasonably required to carry out its proposed mining programme. Any required surface rights not covered by a surface rights agreement must be lawfully reallocated in accordance with the applicable law for the purpose of enabling the permit holder to exercise its rights and obligations under the mining permit.

Mining Permit Rights and Obligations

Article 34 LMM includes provisions concerning mining permit rights and obligations. A mining permit confers on the permit holder the exclusive right to carry out the specified mining operations and related exploration activities, including the right to (Article 34 LMM): (a) enter the permitted area and to take all reasonable measures on or under the surface for the purpose of carrying out the mining programme authorised by the permit; (b) erect the necessary equipment, plant and buildings for the purposes of exploiting, transporting, or processing the mineral recovered; (c) sell or otherwise transfer rights to any mineral recovered.

A mining permit holder must submit to the Commission (Article 34 (4) LMM): (a) within 30 days from the end of each calendar quarter: (i) a report of the volume/tonnage of minerals extracted during that quarter; (ii) a statement showing the amount of the royalty that has been determined to be payable in respect of such quarter, together with all information and calculations relating thereto; and (iii) written proof that such royalties have been paid in the manner and amount required by law; (b) within 120 days from the end of each calendar year. (c) Not later than September 30 of each calendar year, an updated mining programme for the following calendar year.

A mining permit holder must give written notice to the competent authority 3 months in advance, if he intends to suspend production. Such notice must include a detailed justification for such suspension or reduction (Article 35 (1) LMM).

A mining permit holder who wishes to extend his permit, must submit to the competent authority a complete application for such extension in the prescribed format, together with the prescribed fee, and must attach thereto the following (Article 36 (1) LMM): (a) a copy of the mining permit to be extended and any related permit to conduct special operations; (b) an updated mining programme.

20.3.1.3 Authorities

The main responsible authority for mining is the Independent Commission on Mines and Minerals.

ICMM is authorized to regulate mining activities in Kosovo in accordance with the LMM (Article 55 (2)). In the exercise of its responsibilities, ICMM must act in the public interest and independently (Article 55 (3)).

The Mining Inspectorate, which is a Department within ICMM, has the general authority to inspect exploration, mining and special operations conducted by any person (Article 69). The Mining Inspectorate has investigative and enforcement powers, including the power to confiscate mining assets of illegal mining operators (Articles 70 to 72).

20.3.1.4 Fees and Taxes

The ICMM can issue a schedule specifying the amount of the application fee that must be submitted with each type of application. The amount of such fees must be consistent with international best practice. Each applicant must pay the required fee to the competent authority (Article 7 (3) LMM).

The Assembly of Kosovo adopts a schedule of royalties that permit holders are required to pay in connection with the conduct of exploitation activities in Kosovo. Such schedule specifies the formulae to be used in calculating the amount of royalties due (Article 50 LMM).

Where, for any reason, it is not currently possible to determine the amount of royalty due, the Commission may require the permit holder to pay an estimated royalty (Article 52.1).

If a permit holder fails to pay any royalty when due, the competent authority can prohibit him from selling or otherwise disposing of any interest in any concerned mineral or production until all outstanding royalties have been fully paid (Article 53).

Article 54 includes administrative charges and fees. Every permit holder must pay to the competent authority an annual maintenance fee as specified in a schedule to be issued by the Commission.

20.3.2 Additional Legal Basics

Table 77: List of Acts significant for Raw Materials – Kosovo

Name of Law	No. of Law and Year of Issuing
Law on Public-Private-Partnerships and Concessions in Infrastructure and the Procedures for their Award	Law No. 03/L-090
Spatial Planning Law	Law No. 2003 of 14 of 2003
Nature Conservation Law	Law No. 02/L-18 of 2001
Environmental Protection Law	Law No. 03/L-025 of 2009
Water Act	Law No. 24 of 2004
Waste Law	Law No. 02/L-30 of 2006
Law on Foreign Investment	Law No. 02/L-33 of 2005
Law on the Administrative Procedure	Law No. 02-L28
Law on Tax Administration and Procedures	Law No. 2004/48
Law on Noise Protection	Law No. 02-L102
Law on Environmental Impact Assessment	Law No. 03-L-024
Law on Forests in Kosovo	Law No. 3 of 2003
Law on Air Protection	Law No. 30 of 2004
Law on the Activities of Water and Waste Water	Law No. 03/L-086
Law on Cadastre	Law No. 25 of 2003

Environmental Protection Law

This law shall harmonize economical development and social welfare with basic principles for environmental protection according to the concept of sustainable development (Article 1). According to Article 29, mining projects require an environmental impact assessment.

Water Act

This Law regulates issues relating to the management, planning, protection and institutional responsibilities in regard to water and water resources (Article 1). The purpose of this Law is to ensure the development and sustainable use of the water resources necessary for human health, the environment and the socio-economic development of Kosovo; to establish procedures and guiding principles for the optimal allocation of water resources based upon use and purpose (Article 2). According to Article 56 of the Water Act a water permit is required for mining activities and geological works which affect the water regime and the extraction of sand, gravel, stones and clay.

Nature Conservation Law

This law creates the basic requirements for nature conservation and for its sustainable utilization, in particular conservation, renewal and sustainable utilization of nature and the nature renewable resources; restoration of damaged nature conservation zones and of their natural habitats and species; maintenance and restoration of the ecological balance of nature (Article 2). Article 7 includes a strategy and plan of action for nature conservation. The strategy and plan of action on nature conservation, contains measures for sustainable exploitation of natural resources.

21 Latvia

21.1 General Facts

The national territory covers 64 589 km² and the inhabitants are 2.231.503 (2009 estimation). The population density is 36/km². The GDP per capita for 2009 is estimated at $ 14.500.

Constitutional Structure

The form of government is a unitary parliamentary republic. Latvia is divided into 118 municipalities

21.2 Production of Raw Materials

Table 78: Production Data – Latvia (World Mining Data, Weber and Zsak, 2008)

Resource		2002	2003	2004	2005	2006	Change 02/06	Change 05/06
Gypsum	(t)	228.800	265.050	266.240	266.240	230.000	0,52	−14,81

Table 79: Production of aggregates – Latvia (Metric tons), (USGS, 2008)

	2002	2003	2004	2005	2006
Crushed rock	n.a.	n.a.	n.a.	414 305	586 607
Gravel, pebbles, shingle and flint of a kind used for concrete aggregates; for road or for railway and other ballast	n.a.	3.070.709	2.094.017	2.817.287	3.824.965
Limestone	217.074	159.133	225.742	220.000	230.000
Sand and gravel	761.614	1.981.431	1.875.494	1.875.494	2.132.779
Silica sand, industrial	50.000	50.000	50.000	50.000	50.000

21.3 Normative Basics

21.3.1 Primary Legal Basics

The primary legal basic of mineral extraction activity is the Mining Law ("Law on the Subsoil") No. 13 of 1996 as amended by Law No. 321/322 of 2000.

21.3.1.1 General Rules

Table 80: Structure of the Mining Law – Latvia

Section I General Provisions	Article 1 Terms used in the text of the Law	Item 1–26
	Article 2 The implementation of the Law	
	Article 3 Property rights	Item 1–2
Section II Supervision of the Subsoil Fund of the Republic of Latvia	Article 4 Institutions supervising the use of the subsoil fund	Item 1–7
	Article 5 The forms of supervision of the use of the subsoil	Item 1–3
Section III The Use of the Subsoil	Article 6 Basic principles of the use of the subsoil	Item 1–4
	Article 7 Purposes of the use of the subsoil	Item 1–6
	Article 8 Users of the subsoil	Item 1–3
	Article 9 Types and duration of the use of the subsoil	Item 1–4
	Article 10 The procedures of the use of the subsoil	Item 1–10
	Article 11 The use of the subsoil without permits (licences) for the use of the subsoil	Item 1–2
	Article 12 Easement rights in the use of the subsoil	Item 1–5
	Article 13 The rights of the users of the subsoil	Item 1–5
	Article 14 The obligations of the users of the subsoil	Item 1–10

Section IV Protection of the Subsoil	Article 15 Principal requirements for the protection of the subsoil	Item 1–5
	Article 16 Limitation, interruption and termination of the use of the subsoil	Item 1–4
	Article 17 Conditions for new construction in the areas of occurrence of mineral resources	Item 1–2
	Article 18 Control of the use and protection of the subsoil	Item 1–2
Section V Responsibility for Violations During the Use of the Subsoil and Compensation of Damages	Article 19 Responsibility for violations during the use of the subsoil	
	Article 20 Compensation of damages	Item 1–3
	Article 21 The responsibility for the damages inflicted by previous landowners and users of the subsoil	Item 1–3
	Article 22 The information base of the use and the protection of the subsoil	Item 1–5
	Article 23 The basic principles of the use of geological information	Item 1–4
Section VI Final Conditions	Article 24 Resolution of disputes	
Conditions of Transition		Item 1–4
Appendix	List of Common Minerals	Item 1–4

Ownership of minerals rights

The subsoil and all mineral resources therein belong to the land owner (Article 3 ML). According to Article 5 ML the Cabinet of Ministers in the interest of the State has the right to limit the rights of legal and physical entities regarding the land and the subsoil belonging to them by imposing the limitations of the right to use the property.

21.3.1.2 Issuing of Permits

Exploration

Permits for exploration of other minerals are issued by Geological Survey of Latvia (SGSL) based on applications. The procedures for obtaining such permits are described by Regulations of the Cabinet of Ministers No. 239 (1997). For a limited time period, the subsoil may be allocated for exploration purposes up to 5 years.

Extraction

For a limited time period, the subsoil may be allocated: 1) for exploration purposes up to 5 years; 2) for the production of minerals up to 25 years; 3) for exploration and subsequent production of minerals up to 30 years. According to Article 13 ML the operator has the following rights:

1) to use the subsoil for entrepreneurial activities mentioned in the permit;
2) to make use of the products obtained as a result of the use of the subsoil in compliance with the issued permit and the existing legal acts;
3) to make use of the by-products obtained during the production and processing of mineral resources if the permit does not contain limitations;
4) to ask the permit issuing authority to change the conditions contained in it if, during the use of the subsoil, circumstances have been encountered which considerably differ from those mentioned in the permit;
5) to obtain prolongation of the term of the permit for the use of the subsoil or to obtain a new one if the conditions of the previous one have been properly fulfilled and if such is permitted by the agreement with the land owner.

According to Article 14 ML the obligations of the operator are as follows:

1) to comply with legal requirements, standards, norms, permit requirements and other regulations in conjunction with the activities aimed at the use of the subsoil;
2) to comply with the regulations for the production of mineral resources and occupational safety regulations;
3) during exploration of the subsoil, to produce geological documentation and to safeguard its safe storage;
4) to transfer to SGSL geological information and data in compliance with the terms and deadlines contained in the permit as well as the data about the reserves of mineral resources and their components;
5) to submit necessary reports about the use of the subsoil in compliance with the procedures contained in legal acts;
6) to comply with standards, norms and regulations safeguarding the protection of the environment and cultural monuments, land transformation, the protection of structures and other objects and prevent adverse effects on them as a result of the use of the subsoil.

Protection of subsoil (Article 15 ML)

The principal requirements for the protection of the subsoil are as follows: 1) complete and comprehensive exploration of the subsoil; 2) rational production of mineral resources and use of by-products found in the deposits; 3) the use of the subsoil which does not lead to adverse effects on the reserves of mineral resources and useful properties of the subsoil (Article 15 ML).

21.3.1.3 Authorities

The main responsible authority for mining is the Ministry of Environmental Protection and Regional Development (MEPRD).

The permits for the use of the subsoil are issued by the local authorities in cases mentioned in Article 4 (5) ML (common minerals); the State Geological Survey (under administration of MEPRD) in all other cases.

The supervision of the use of the subsoil is carried out, on behalf of the State, notwithstanding the type of property (owner), in compliance with procedures contained in the administrative acts, by the Ministry of Environmental Protection and Regional Development (MEPRD) and institutions within its structure, under its administration and supervision; the Ministry of Economy (ME); the local authorities of parishes and towns (cities).

21.3.1.4 Fees and Taxes

Permits (licences) for the use of minerals are issued for a fee. The size of the fees and the procedures of their payment are determined by the Cabinet of Ministers.[1]

21.3.2 Additional Legal Basics

Table 81: List of Acts significant for raw materials – Latvia

Name of Law	Year of Issue
The Law "On Natural Resource Tax"	1995
The Law "On Environmental protection"	1991
The Law "On Environmental Impact Assessment"	1998
The Law "On Spatial Development Planning"	1999
The Law "On Pollution"	2002
Administrative Procedure Law	2001
Civil Procedure Law	1998
Land Register Law	

---

1 Chodak, M. (2004), Country Report Latvia, in: Department of Mining and Tunnelling, lc.

Name of Law	Year of Issue
Law On Land Use and Land Survey	1993 last amended 2007
The Civil Law	1938 last amended 1992
Regulations of the Use of Minerals, Deposits and Subsoil Areas of State Importance	No. 238 of 1997
Regulations on the use of subsoil	No. 239 of 1997
The Provisions of the Use of Minerals, Deposits and Subsoil Areas of State Importance	No. 307 of 2000
On Methods and Tariffs (Rates) for Compensation of Losses Caused to Subsoil	No. 298
Work safety requirements during exploration for and production of minerals	No. 253 of 2002
Procedures for Application of Certain Provisions of the Law On Natural Resources Tax	No. 356 of 2000
Procedures of calculation and payment of the natural resource tax	No. 244 of 2002
Procedures for Environmental Impact Assessment	No. 213 of 1999
Regulations on Territorial Planning	No. 423 of 2000

Environmental Assessment Law

The purpose of the Environmental Assessment Law is to prevent or reduce the negative impact of the implementation of the intended activities of natural persons and legal persons or of a planning document thereof on the environment (Article 2). Projects designed for the extraction of mineral resources in previously unused mineral resource deposits, the area of which is larger than 10 hectares or in previously unused peat deposits, the area of which is larger than 100 hectares, regardless of the area intended for processing (Annex 1).

22 Lithuania

22.1 General Facts

The national territory covers 65 200 km² and the inhabitants are 3.555.179 (2009 estimation). The population density is 52/km². GDP per capita for 2009 is estimated at $ 11.172.

Constitutional Structure

The form of government is a semi-presidential republic. It is divided into 10 counties that are further divided into 60 municipalities which consist of over 500 elderships. Counties are ruled by a county governor, appointed by the central government. Municipalities are the most important administrative units; they have their own elected governments. Elderships are engaged in local public services.

22.2 Production of Raw Materials

Table 82: Production Data – Lithuania (World Mining Data, Weber and Zsak, 2008)

Resource		2002	2003	2004	2005	2006	Change 02/06	Change 05/06
Oil	(t)	433.727	382.100	360.000	216.100	180.900	−58,29	−58,29

Table 83: Production of aggregates – Lithuania (Metric tons unless otherwise specified), (USGS, 2008)

	2002	2003	2004	2005	2006
Clays	n.a.	240.800	228.100	289.500	385.300
Granules, chippings and powder of stones, excluding marble	n.a.	n.a.	n.a.	4.316	10.390
Limestone	984.300	944.600	1.385.600	1.242.200	1.776.300

	2002	2003	2004	2005	2006
Marble granules, chippings and powder	n.a.	n.a.	n.a.	666	1.167
Sand and gravel:					
Construction sands	n.a.	n.a.	n.a.	3.689.217	4.342.743
Gravel, pebbles, shingle and flint	n.a.	n.a.	n.a.	3.345.185	3.290.568
Silica sand, industrial	63.000	49.700	58.300	46.500	42.600

22.3 Normative Basics

22.3.1 Primary Legal Basics

The primary legal basics of mineral extraction activity is the Underground Law No. VIII-573 of 1995.

The Law on State supervision of precious metals and gems No. I-996 of 1995 regulates the procedure for assaying precious metals and gems, as well as the articles thereof, establish the legal basis of the activities of institutions which exercise the state supervision of precious metals and gems, as well as the articles thereof, and obligatory requirements for economic entities which engage in commercial-economic activities relating to precious metals (e. g. precious metals as gold, silver, metals of the platinum group (ruthenium, rhodium, palladium, osmium and iridium) and gems.

22.3.1.1 General Rules

Table 84: Structure of the Mining Law – Lituania

Chapter I General Provisions	Article 1. The Purpose of the Underground Law	
	Article 2. Ownership of the Underground	
	Article 3. Main Definitions	
Chapter II State Regulation of the Utilisation, Protection and	Article 4. General Competence of the Government of the Republic of Lithuania	Section 1–3
	Article 5. Competence of Special Public Institutions	Section 1–3

Control of the Underground	Article 6. Competence of the Governor of the County and Municipal Institutions	Section 1–2
Chapter III **Investigations of the Underground**	Article 7. Permit to Carry out the Investigation of the Underground	Section 1–2
	Article 8. Registration of the Investigations of the Underground	Section 1–2
	Article 9. Conditions of Investigation of the Underground	Section 1–4
	Article 10. The Revocation of the Permit for the Investigation of the Underground	Section 1–2
	Article 11. State Geological Surveys	Section 1–2
Chapter IV **Exploitation of Underground Resources or Caves**	Article 12. The Procedure for Exploitation of Underground Resources and Caves	Section 1–2
	Article 13. The Right to Exploit Underground Resources or Caves	Section 1–6
	Article 14. Granting Permits to Exploit Underground Resources and Caves	Section 1–6
	Article 15. The Plan for Exploitation	Section 1–2
	Article 16. The Conditions for Exploitation of the Underground Resources	Section 1–6
	Article 17. Allotting of a Piece of Land	
	Article 18. Expiration of the Validity of the Permit to Exploit the Underground Resources and Caves	Section 1–3
	Article 19. The Revocation of the Licence to Exploit the Underground Resources and Caves	Section 1–3
Chapter V **The Protection of the Underground**	Article 20. Measures of the Protection of the Underground	
	Article 21. Territorial Planning	Section 1–3
	Article 22. The Monitoring of the State of the Underground	Section 1–3
	Article 23. The Protection and Use of the Underground in the Protected Areas	
Chapter VI **Data about the Underground**	Article 24. Required Providing of Data on the Underground	Section 1–2
	Article 25. State Geological Information System	Section 1–3
	Article 26. The Utilisation of the Data about the Underground	Section 1–8

Chapter VII Responsibility for the Violation of the Underground Law and the Settlement of Disputes	Article 27. Responsibility for the Violation of the Underground Law	
	Article 28. Claims for the Damage Caused by Illegal Activities	Section 1–2
	Article 29. Settlement of Disputes	
Chapter VIII International Relations	Article 30. Influence on the Environment of other States	
	Article 31. International Co-operation	
	Article 32. International Agreements	
Chapter IX Final Provisions		

Minerals classification

Mineral resources are classified in non-metallic mineral resources; metal ores; valuable minerals and hydrocarbons.

Ownership of mineral rights

The mineral resource is the exclusive ownership of the State. The basis of the exploitation of mineral resources is the right to exploitation, which can be granted to legal and natural persons by the competent authority (Article 2 ML).

Acquiring mineral rights

Mineral resources can be exploited by the legal and natural persons who have acquired a permit issued by the competent authority, and who have concluded an exploitation contract with it. A licence to exploit the mineral resources grants to the person who holds it, the exclusive right to exploit the types of mineral resources or caves, which are indicated in the licence, in the specified area, during a set period of time, in conformity with the terms of the exploitation contract (Article 13).

22.3.1.2 Issuing of Permits

Prospection and exploration

Prospection of the mineral deposits can be carried out by legal and natural persons, having a permit for this kind of economic activity. The exploration permits for legal and natural persons of the Republic of Lithuania and foreign countries are issued by the Geological Survey of Lithuania (Article 7 ML).

Exploration of the mineral deposits of all types must be registered by the Geological Survey. Prior to their commencement, direct investigations of the mineral deposits must be reported to the board of the municipality, on the territory whereof the exploration is planned, and to the land survey of the administration of the governor of the county (Article 8 ML).

Conditions of exploration of the mineral deposits are: The utilisation of mineral resources in the course of the investigation of the mineral deposits can be carried out only in ways provided for in the work plan. The Geological Survey of Lithuania may instruct the permit holder to carry out additional investigations, related with his work (Article 9 ML).

Exploitation

1. Mineral resources can be exploited by the legal and natural persons of the Republic of Lithuania and foreign countries, who have acquired a permit issued by the competent authority, and who have concluded an exploitation contract with it (Article 13 ML).
2. A permit to exploit the mineral resources grants to the person who holds it, the exclusive right to exploit the types of mineral resources which are indicated in the permit, in the specified area, during a set period of time, in conformity with the terms of the exploitation contract.
3. Investigation (i. e. exploration) of the mineral resources may be provided for in the permit to exploit mineral resources, specifying the area and period of these investigations, and establishing their terms in the exploitation contract. In this case, the permit must confer the right to exploit also the newly discovered or additionally investigated resources.
4. Permits to exploit the mineral resources specified in Article 14 (1) ML will be issued on a competitive bidding.
5. Exploration and exploitation of the mineral resources of other types can be carried out in the same area and at the same time on the basis of a separate permit, provided that it does not impede the activities of other persons, who have earlier acquired the permit.

According to Article 14 (5) ML all the exploitation permits must be registered in the Geological Survey of Lithuania. The State institution granting permits to exploit mineral resources shall inform the municipality, the Land survey of the administration of the governor of the county and the public about these permits, before starting the activities provided for them (6).

The extraction plan

According to Article 15 (1) ML exploitation of mineral resources is possible only on the basis of an extraction plan which is co-ordinated with the governor of the county and approved by the Ministry of the Environmental Protection. The following must be provided for in the extraction plan: measures for recultivation of land, as well as necessary measures for restoration of other elements of the environment; measures for the protection of mineral resources, left in the deposit, from the exhaustion and decrease in quality, when the exploitation of the deposit is temporarily or completely terminated (Article 15 (2) ML). In course of the first five years of exploitation of the mineral resources, the operator must accumulate the funds necessary for the fulfilment of measures specified in Article 15 (2) ML and guarantee the use of the funds for those purposes.

According to Article 16 ML the conditions for exploitation of the mineral resources are:

1. Mineral resources can be exploited only after appraisal whereof and upon having evaluated the influence of their extraction on the environment.
2. Mineral resources must be exploited in complex or protecting not utilised resources, being in the same deposit.
3. Mineral resources must be exploited rationally and only for the purposes, indicated in the permit.
4. The limits for the amount of the mineral resources to be extracted, sold or exported must be indicated in the exploitation contract.
5. In course of the exploitation of the deposit, it is obligatory to monitor the state of the resources, to predict changes in its quantity and quality and the influence of the exploitation on the environment, and carry out an accounting of the deposits which are extracted and remaining in the deposit.

The Ministry of Environment Protection and the Geological Survey of Lithuania must be provided with the data of these observations. The methods and the amount of the monitoring must be provided for in the plan of exploitation and financed by the permit holder.

The permit to exploit the mineral resources expires when: 1) the period of the validity of the permit expires; and 2) the object of the exploitation is exhausted (Article 18 ML). The exploitation permit is revoked when the parties, concluding the exploitation contract, do not agree on the terms of the contract (Article 19 ML).

22.3.1.3 Authorities

The main responsible authority for mining is the Minister of Environment. According to Article 14 ML

1. The Government grants permits for exploitation of metallic minerals, monomineral quartz sand;
3. The Geological Survey of Lithuania, upon co-ordinating with the Ministry of the Environmental Protection and the administration of the governor of the county, grants permits to exploit the mineral resources not listed in (1) and (2) of Article 14 ML.

22.3.1.4 Fees and Taxes

The cost of permit for exploration is 150 Lt (44 Euro). The cost of permit to use mineral resources depends on the sort of mineral resources: metal ores- 2500 Lt (724 Euro); other mineral resources 200 Lt (58 Euro).[1]

There is a number of different taxes fees and compensations existing in Lithuania. As a compensation for exploitation of mineral resources an operator must pay the natural resource tax. In case of mineral resources extraction, the amount of the tax depends on the volume of extracted mineral resource. The natural resource tax is a part of the taxation system of Lithuania and is paid to the state budget.

An operator must pay pollution taxes for emission of harmful substances into and storage and of wastes.

Allotting and using the land plot for industrial (extracting) purposes is connected with paying the land tax.

Furthermore, extracting firms as each entity providing economic activity must pay other commercial taxes.

---

1 Uberman, R., Ostrega, A. (2004), Country Report Lituania, in: Department of Mining and Tunnelling, lc.

22.3.2 Additional Legal Basics

Table 85: List of Acts significant for raw materials – Lithuania

Name of the Law	No. of Law and Year of Issuing
Law on Environmental Protection	Law 1992 as amended by Law No. I-2322 of 1996
Law on Forestry	1994
Law on Protected Areas	No. I-301 of 1993 last amended IX-628 of 2001
Law on Water	No. VIII – 474 of 1997
Resolution on Approving Permit Granting Procedure to Exploit the Resources of Minerals	2002

Law on Environmental Protection

This Law regulates public relations in the environmental protection field, define the main rights and duties of legal and natural persons preserving biological diversity characteristic to the Republic of Lithuania, ecological systems and landscape, ensuring healthy and clean environment, rational use of natural resources in the Republic of Lithuania, its territorial waters, continental shelf and economic zone (Article 2). Natural resources must be utilized in a rational and composite way taking into consideration the possibilities of preservation and renewal of nature (Article 4).

Water Act

This Law regulates the ownership of the internal bodies of water of the Republic of Lithuania, the management, use and protection of their water resources, relations between the owners and users of water bodies and the rights and obligations of legal and natural persons using internal bodies of water and their resources (Article 1). Mining activities shall be organised and conducted in such a way as to minimise the adverse effects on water quality, flora and fauna as well as the stability of the banks and hydro technological facilities (Article 34).

23 Luxembourg

23.1 General Facts

The national territory covers 2,586,4 km² and the inhabitants are 493.500 (2009 estimation). The population density is 186/km². The GDP per capita for 2009 is estimated at $ 104.512.

Constitutional Structure

The form of government is a parliamentary democracy and Constitutional Grand Duchy. It is divided into 3 districts (Diekirch, Grevenmacher, Luxembourg), subdivided into 12 cantons and then 116 communes.

23.2 Normative Basics

23.2.1 Primary Legal Basics

The primary legal basics of mineral extraction activity is Law of May 1990 related to the control of dangerous, polluted and noxious installations.

The basics of the legislation related to mineral extraction go back to Mining Laws from the 19th century (e. g. Law of 21 April 1810, of 14 October of 1842 and of 30 April 1890). More recently, the Law of May 1990 related to the control of dangerous, polluted and noxious installations has become the key legislation for new projects, including quarry operations.

23.2.1.1 General Rules

Ownership of mineral rights

Mineral resources deeper than 6 m are owned by the State and subjected to the payment of royalties. The owner of the land owns mineral resources near the surface. As the land owner owns the industrial minerals and building materials, one should come to an agreement with the land owner.

23.2.1.2 Issuing of Permits[1]

Prior authorisation is not required for borehole investigation, but may be required for trial excavations or other forms of exploration (e. g. trenching).

For the extraction of mineral resources, the disposal of waste and the backfilling and restoration of the excavations, one must obtain a permit at national level. The Law of May 1990, related to dangerous, polluted and noxious installation, is the key legislation. No mineral activities are automatically authorised and all are subject to the same procedure. An environmental impact study may be required. The applicant has first to provide a summary assessment and based on this, the competent authorities decide if a detailed environmental impact study is required.

Mineral operators normally discuss their proposal with the appropriate government department (e. g. ITM, Inspectorate of Works and Mines and Ministry of the Environment) prior to making an application. There must be a public notice of the fact that an application for exploitation has been submitted, e. g. a notice must be simultaneously posted at the town hall and in a very visible location near the proposed site. In areas with more than 5.000 inhabitants, the notice must also be published in four daily newspapers. After a period of 15 days, the written remarks are collected and a public hearing is held in the community concerned. Any interested party has the right to participate, including citizens and environmental groups from neighbouring countries. The findings of the hearing are summarised in a written report and submitted to the competent authority. In global terms, the criteria used to evaluate an application are environmental protection, health and safety of the workforce, and protection of the rights of third parties.

Restoration

The rehabilitation of an extraction site forms part of the environmental impact study. As for other items, rules for the rehabilitation can be attached to the conditions of authorisation. The operator is responsible for the costs of rehabilitation and must provide a bank guarantee for these costs.

23.2.1.3 Authorities

Authorisation for mineral extraction is granted by the National Government. The Inspectorate of Works and Mining issues the permit, following consultation with the Ministry of Environment. ITM belongs to the Ministry of

---

[1] Vervoort, A. (2004), Country Report Luxembourg, in: Department of Mining and Tunnelling, lc.

work and employment, and its organisation is defined by the Law of 4 April 1974.

The monitoring of the extraction is provided by the Law of May 1990, related to the control of dangerous, polluted and noxious installations. This enables the competent authority (e. g. Ministry of the Environment) to review the operator's compliance with the requirements and conditions concerning environmental protection, as stipulated in the permit. The conditions of authorisation can be revised, if negative effects occur which had not been foreseen. Art. 22 of the Law of May 1990 provides the possibility to stop immediately the extraction, if the conditions attached to the permit are not fulfilled.[2]

23.2.1.4 Fees and Taxes

23.2.2 Additional Legal Basics[3]

Table 86: List of Acts significant for raw materials – Luxembourg

Name of Law	Year of Issuing
Law related to the control of dangerous, polluted and noxious installations	1990
Law related to the protection of nature and of natural resources	1982
Law related to the various aspects of land use	1999
Law related to the safety and health of workers	1994

Mining Laws from the 19th century (e. g. Law of 21 April 1810, of 14 October of 1842 and of 30 April 1890).

The Law of 1990 related to the control of dangerous, polluted and noxious installations has become the key legislation for new projects, including quarry operations. It sets out a discrete authorisation procedure, based on the concept of Best Available Technology (BAT). The legislation also makes provisions for greater public involvement in the decision-making process and increases owner liability. The Law of 1982 covers the protection of nature and of natural resources. By referring to the Law of 1974, "Green zones" are being defined. In such zones, the start of mining or quarrying operations is subjected to the authorisation by the Minister responsible for Water and Forest (Chap. 2, Art. 4). The Minister can order an impact study, prior to his decision.

For the use of explosives, the Inspectorate of Works and Mining ("Inspection du Travail et des Mines") has published safety regulations on 20 August

---

2   Ibidem.
3   Ibidem.

2001. These regulations refer to the Law of 17 June 1994 and of 4 November 1994 on the safety and health of workers.

Some recent laws explicitly eliminate the application of mineral extraction, if a specific "Mining" Law is covering it. This is for example the case for the Law of 17 June 1994 on the prevention and handling of waste (Chapter I, Art. 2.f). In a similar way, the Order of 17 July 2000 on the danger linked to major accidents excludes the activities of mines, quarries and drilling (Art. 4.d).

24 Macedonia
Former Yugoslav Republic of Macedonia – FYROM

24.1 General Facts

The national territory covers 25 713 km² with an estimated total population until 2009 of 2,114,550 inhabitants, population density is 82,2/km². The GDP per capita for 2009 is estimated at $ 4.482.

Constitutional Structure

The form of government is a parliamentary republic. It is divided into 84 municipalities (since August 2004).

24.2 Production of Raw Materials

Table 87: Production Data – Macedonia (World Mining Data, Weber and Zsak, 2008)

Resources		2002	2003	2004	2005	2006	Change 02/06	Change 05/06
Nickel	(t)	5.100	5.500	5.300	8.100	10.900	1,3.73	34,57
Cadmium	(t)	111	75	0	0	0	–1,0.00	
Copper	(t)	5.600	700	0	6.000	8.400	50,00	40,00
Lead	(t)	15.000	2.700	800	800	15.600	4,00	1,50,00
Zinc	(t)	10.000	4.000	0	0	21.700	1,7.00	
Gold	(kg)	500	400	0	0	0	–1,0.00	
Silver	(kg)	8.000	1.000	0	0	10.000	25,00	
Bentonite	(t)	3.000	3.500	3.700	3.900	4.500	50,00	15,38
Feldspar	(t)	21.000	21.000	22.000	23.000	20.000	–4,76	–13,04
Talc	(t)	800	800	900	1.000	500	–37,50	–50,00
Lignite	(t)	8.600.000	7.492.000	7.300.000	6.879.726	5.874.000	–31,70	–14,62

Table 88: Production of bentonite, gypsum, lime and construction minerals – Macedonia. (Metric tons unless otherwise specified), (USGS, 2008)

		2002	2003	2004	2005	2006
Clays, bentonite		25.000	25.000	25.000	25.000	25.000
Gypsum, crude		150.000	150.000	150.000	190.232	267.760
Lime		500	500	500	500	500
Sand and gravel, excluding glass sand	1000 m³	100	100	100	100	100
Stone, excluding quartz and quartzite:						
Dimension, crude	m²	150.000	150.000	150.000	150.000	150.000
Ornamental	1.000 m³	300	300	300	300	300
Crushed and broken, n. e. s.	m³	5.000	5.000	5.000	5.000	5.000
Other	1.000 metric tons	25.000	25.000	25.000	25.000	25.000

24.3 Normative Basics

24.3.1 Primary Legal Basics

The primary legal basics of mineral extraction activity is the Mining Law ("Law on Minerals Raw Materials").

24.3.1.1 General Rules

Table 89: Structure of the Mining Law – Macedonia

Part I General Provisions	Article 1 Subject and goal of the law	Item 1–2
	Article 2 Definitions	
	Article 3 Types of mineral ores (raw materials)	Item 1–2

Part II Principal and Detailed Geological Explorations		Article 4 Geological-mining Works	
		Article 5 Strategy of Geological Explorations, Sustainable Recovery and Mining of Mineral Ores	Item 1–4
		Article 6 Conditions of Performing Principal Geological Explorations	Item 1–5
	Chapter I Principal Geological Researches	Article 7 Goal of Primary Geological Explorations	Item 1–3
		Article 8 Annual Program on Primary Geological Explorations	Item 1–3
		Article 9 Geological Documentation for Performing Primary Geological Explorations	Item 1–4
	Chapter 2 Detailed Geological Explorations	Article 10 Permit for Performing of Geological Explorations	Item 1–2
		Article 11 Application for Issue of Permit to Perform Detailed Geological Explorations	Item 1–3
		Article 12 Priority Right to Issue Permit for Performing of Detailed Geological Explorations	Item 1–3
		Article 13 Procedure to Issue Permit to Perform Detailed Geological Explorations	Item 1–3
		Article 14 Area to Perform Detailed Geological Explorations	Item 1–2

		Article 15 Validity Period of Permit to Perform Detailed Geological Explorations	
		Article 16 Rejection of Application to Issue Permit for Detailed Geological Explorations	Item 1–2
		Article 17 Content of Permit to Execute Detailed Geological Explorations	Item 1–15
		Article 18 Conditions for Extension of Permit to Perform Geological Explorations	Item 1–4
		Article 19 Obligations of Permit Holder related to Performing of Detailed Geological Explorations	Item 1–3
		Article 20 Deprive form Permit to Perform Detailed Geological Explorations	Item 1–3
		Article 21 Ownership of Results from Detailed Geological Explorations	Item 1–5
		Article 22 Geologic Documentation relate to Performing of Detailed Geologic Explorations	Item 1–3
Part III Mining of Mineral Raw Materials	Chapter I Concession for Mining of Mineral Raw Materials	Article 23 Concession for Mining of Mineral Ores	Item 1–3
		Article 24 Subject to Assignment of Concession	
		Article 25 Area for Mining of Mineral raw materials	Item 1–5

Article 26 Period of Concession Validity	Item 1–3
Article 27 Method of Granting Concession	
Article 28 Granting of Concession by Invitation to Bid Notice	Item 1–4
Article 29 Granting of Concession for Mining of Mineral Ores upon Applications by Interested Entities	Item 1–6
Article 30 Rejection of Application for Concession on Mining	Item 1–2
Article 31 Restriction in Granting of Concession on Mining	
Article 32 Decision related to Granting of Concession on Mining	Item 1–2
Article 33 Contract for Concession on Mining of Mineral Ores	Item 1–3
Article 34 Assignment of Concession on Mining	Item 1–5
Article 35 Termination of Concession on Mining	Item 1–5
Article 36 Termination of Concession on Mining of Mineral Ores after the Expiry	
Article 37 Prior Purchase of Concession on Mining of Mineral Ores in Public Interest	Item 1–2

	Article 38 Termination of Concession on Mining of Mineral raw Materials due to Prior of Time Depletion	
	Article 39 One-sided Termination of Contract on Concession related to Mining of Mineral raw Materials	Item 1–6
	Article 40 Concession related to Mining of Another Type of Mineral Ores (Raw Materials) on the Same Area	Item 1–2
Chapter2 Permit Related to Mineral Raw Materials Mining	Article 41 Permit related to Mineral Ore Mining	Item 1–3
	Article 42 Application for Issue of Permit related to Mining of Mineral raw Materials	Item 1–3
	Article 43 Content of Permit related to Mining of Mineral Ores	Item 1–2
	Article 44 Rejection of Application related to Issue of Permit on Mining of Mineral Ores	Item 1–2
	Article 45 Termination of Permit related to Mining of Mineral Ores	Item 1–3
	Article 46 Deprive from Permit related to Mining of Mineral Ores	Item 1–2
	Article 47 Permit Perform Mining Works according to Mining Design	Item 1–8

		Article 48 Termination and Deprive from the License related to Executing Mining Works according to the Supplemental Design	Item 1–2
		Article 49 Mining Designs	Item 1–4
	Chapter 3 Performance of Mining Works	Article 50 Terms and Conditions to Do Mining Works	Item 1–2
		Article 51 Concessionaire's Obligations in Doing the Mining Works related to Mineral Ores Mining	Item 1–5
		Article 52 Mining Surveys and Mining Plans	Item 1–2
		Article 53 Works Managing and Supervision	Item 1–3
		Article 54 Temporary Suspension or Permanent Interruption of Mining Works	Item 1–4
	Chapter 4 Technical Inspection of Constructed Mine Facilities	Article 55 Technical Inspection of Mine Facilities	Item 1–7
		Article 56 Commissioning without a Certificate of Use for the Mine Facility	Item 1–3
Part IV Technical Preparation, Designers and Reviewers of Geological Documentation and Procedure of Review		Article 57 Educational skills	Item 1–6

24.3 Normative Basics

		Article 58 Entities for Completion of Technical Documentation	Item 1–3
		Article 59 Entities for Preparation of Mining Documentation	Item 1–3
		Article 60 Expert Assessment (review) of Mining Designs	Item 1–7
		Article 61 Review Report and Review Clause	Item 1–5
Part V Payments for Detailed Geological Researches and Concessions related to Mining of Mineral Raw Materials		Article 62 Payments for Detailed Geological Researches	Item 1–2
		Article 64 Determination of Level and Method of Payment of Fee	Item 1–3
Part VI Limitation of the Right to Property		Article 65 Limitation of Ownership Right	
		Article 66 Public Interest	
Part VII Safety at Work		Article 67 Safety at Work in Performing Geological Researches and Mining of Mineral Ores	Item 1–19
		Article 68 Obligation to Getting Familiar	Item 1–5
		Article 69 Obligation for Notification of accidents	
Part VIII Environmental Protection and Cover of Loss		Article 70 Environment Protection	Item 1–2
		Article 71 Cover of Loss (Damage)	

	Article 72 Reclaiming and Removal of Consequences	Item 1–3
	Article 73 Interventions in the Concession Area	Item 1–2
Part IX Supervision	Article 74 Supervision	Item 1–4
Part X Penalty Provisions	Article 75	Item 1–5
	Article 76	Item 1–4
	Article 77	Item 1–4
	Article 78	Item 1–2
Part XI Provisional and Final Provisions	Article 79 Procedures Commenced Prior to Entering into Force of this Law	Item 1–3
	Article 80 Validity of Provisions	Item 1–2
	Article 81 Repealing of law	
	Article 82 Entering into Force of Law	

The types of mineral raw materials according to Article 3 ML:

1) energy mineral raw materials: all types of fossil coals, hydrocarbons in solid, liquid, and gas state, all kinds of bituminized and greasy rocks and other gases that are found on the earth;
2) metallic mineral raw materials and/or raw materials from which metals or their compounds can be produced;
3) non-metallic mineral ores, graphite, sulphur, magnesite, fluor spar, barytes, asbestos, mica, phosphates, gypsum, calcite, chalk, bentonite clay, quartz, quartz sand, kaolin, ceramic and fireproof clay, feldspar, talc, tuff, raw materials for cement and lime production and carbonates and silicate raw materials for industrial processing;
4) architectonic-building stone;
5) building-technical stone, building sand and groove and brick clay.

Ownership of mineral rights

Mineral resources are goods of general interest in property of the Republic of Macedonia regardless of the property of the land it is located on (Article 3 (1) ML).

Acquiring mineral rights

The right to concession for mining of mineral ores can be acquired by obtaining a concession. The concession grants the Government (licensor). The right to granting a concession has any legal and natural entity including also the foreign natural entities with the subsidiaries registered in the Central Register Office of the Republic of Macedonia (Article 23 ML). The government can assign the concessionaire the mineral raw materials in property in a certain period of time (Article 24 ML).

24.3.1.2 Issuing of Permits

Exploration

Principal and detailed explorations can be performed by legal and natural entities with the branches registered in the Central Register Office of the Republic of Macedonia for doing that business (Article 6 (1) ML).

The **primary geological exploration** (i. e. prospection) is an activity of a public interest (Article 7 (1) ML). It provides principal geological data used for mining (2). The primary geological exploration must be implemented in compliance with the annual program on principal geological researches laid down in Article 5 of this law (3).

According to Article 8 (1) ML the funds for financing the annual program under paragraph (1) of this Article are provided from the budget of the Republic of Macedonia (2).

According to Article 9 (1) ML the geological documentation related to performing the primary geological explorations includes: 1) a project study on primary geological explorations; 2) a study report on primary geological explorations; 3) a report on the completed primary geological explorations; 4) a study report on the assessment of the impact on environment.

The right to **detailed geological explorations** can be acquired by obtaining a permit (Article 10 (1) ML). The permit for explorations is issued by the authority of governmental administration competent for doing the works in the field of mineral ores (2).

The application for issuing a permit to perform explorations must state the type of mineral raw material, the location filing the application for and the principal data of the permit applicant (2). The application must be complemented with (3):

1) a topographic map in scale of 1:25,000 or 1:50,000 on which the coordinates of the boundary points of the location are placed for the purpose of which the performance of detailed geological explorations are requested and a geodetic study report issued by the authorized institution together with the numeric data and cadastre plans in scale of 1:25.000;
2) a detailed geological exploration design;
3) a proof for provided financial resources for performing exploration in compliance with the design on detailed geological explorations, and
4) a study report on assessment of environment impact.

Exploitation

According to Article 25 ML a concession for mining of mineral raw materials must be granted to a certain area defined by coordinates on a topographic on the basis of the included study of concession viability and/or the concession design and study report related to the executed detailed exploration (1). The concession for mining of mineral ores may be carried out on a defined area (i. e. "mine field"), defined by the final and additional mining design on mining of mineral ores (3).

Concession for mining of mineral ores are granted for a period of 30 years with the possibility to be extended for another period until the exhaustion of the mineral ores for which the concession has been granted, but not longer than 30 years (Article 26 (1) ML). The request for concession extension must be submitted at least two years prior to the expiry of the period for which the concession has been granted (2). The granting of concession is based on a public invitation to bids or upon a request of the interested entities meeting the conditions of this or of another law (Article 27 ML).

Granting of concession upon application (Article 29 ML)

The concession will be granted upon a request by the interested entity only for the mineral ores laid down in Article 3 (2) ML (i. e. metallic ores). The request for concession related to mineral raw materials mining must be submitted to the competent authority for carrying out the activities in the field of mineral raw materials. According to Article 29 (3) ML the applicant must include the following documents:

a feasibility study in relation to the requested concession and concession project, respectively;

a topographic map in scale 1:25,000 with coordinates of border points in the location of a defined area;

a study report on the detailed geological explorations made or an evidence related to the right to ownership or using of the results from the study report with respect to the detailed geological researches made in the requested area for mining, and

a decision on approval of the environment impact assessment study or a decision on the approval of the study report related to the environment impact assessment.

The decision related to granting a concession must include (Article 32 (1) ML):

- the type of mineral raw materials;
- the surface area on which concession on mining is granted, defined with coordinates;
- the time of validity of concession on mining;
- the obligation for payment of concession fee;
- obligations of concession holder in terms of rehabilitation and reclaiming of land degraded from the mining activities.

The decision related to granting a concession must be published in the Official Gazette of the Republic of Macedonia.

Termination of Concession on Mining (Article 35 ML):

The concession on mining terminates in the following cases:

1) Expiration of the validity time of concession related to mining of mineral raw materials determined in the decision on granting concession and in the contract on concession;
2) prior to the expiration time purchase of concession on mining in public interest;
3) prior to the expiration time depletion of mineral raw minerals;

Terms and conditions of mining works (Article 50 ML)

Mining works may execute legal and natural entities being registered in the Central Register Office and hold a permit on doing mining works as well as the foreign legal entities with their branches registered for that business in the Central Register Office of the Republic of Macedonia, and which meet the conditions prescribed by this or another law.

24.3.1.3 Authorities

The main responsible authority for mining is the Minister of Economy. Article 55 ML regulates the inspection of mine facilities. The mine facility may be used after the completion of technical inspection and on the basis of a certificate of use issued (1). The technical inspection of the mine facility is carried out to constructed mine facility in accordance with the final or additional mine design and upon a request of the concessionaire, the commission established by the minister managing the governmental authorities competent for the performance of the works in the field of mineral raw materials (2). The manner of carrying out the technical inspection as referred to (2) of this Article including the level of the costs for the technical inspections prescribes the Government dependant on the complexity of works (3).

24.3.1.4 Fees and Taxes

Payments for detailed geological researches (Article 62 ML)

The holder of permits related to performing of detailed geological explorations of mineral ores must make a straight payment for the area depending on the type of mineral raw materials defined in the permit on executing detailed geological researches. The mining concessionaire is obliged to pay: 1) an annual payment for the area use allocated with the concession for mining, and depending on the type of mineral raw materials that is subject to mining, and 2) payment for mining of mineral raw materials subject to concession.

Determination of level and method of payment (Article 64 ML)

The Government must adopt a tariff structure determining the level of fees with respect to performing detailed exploration and concessions for mining of mineral raw materials referred to in Articles 62 and 63 and depending on the type, quantity and quality of mineral raw materials (1). The funds charged under paragraph (1) of this Article must be paid on the relevant account for payment in the framework of the treasury account (2). The funds charged under paragraphs (1) and (2) of this article:

- 60% are a revenue to the Budget of the Republic of Macedonia (FYROM), and

- 40% are revenue to the Budget of municipality in the territory of which the concession activity is performed (3).

24.3.2 Additional Legal Basics

Table 90: List of Acts significant for raw materials – Macedonia (FYROM)

Name of Law	No. of Law and Year of Issuing
Law on General Administrative Procedure	2005
Law on Environment	Law No. 53 of 2005
Law on Nature Protection	Law No. 67 of 2004
Law on Ambient Air Quality	Law No. 67 of 2004
Law on Waste Management	Law No. 6 of 2004
Law on Administrative Fees	Law No. 17 of 1993
Law on Leasing	Law No. 49 of 2003
Foreign Trade Law	Law No. 45 of 2002
Law on Technological Industrial Development Zones	Law No. 14 of 2007
Law on Trade	Law No. 15 of 2004
Company Law	Law No. 28 of 2004
Customs Tariff Law	Law No. 23 of 2003
Profit Tax Law	Law No. 27 of 2006

Law on Environment

The Law on Environment regulates the rights and the responsibilities of the Republic of Macedonia, municipalities, the City of Skopje and the municipalities of the City of Skopje as well as the rights and the responsibilities of legal entities and natural persons, in the provision of conditions required to ensure protection and improvement of the environment (Article 1). Environmental Impact Assessment study related to mining is required under certain conditions (Article 65).

Law on Nature Protection

This Law regulates the nature protection by protecting the biological and landscape diversity, and the protection of the natural heritage, in protected areas and outside of protected areas (Article 1). Article 22 covers aspects regarding natural resources management.

25 Malta

25.1 General Facts

The national territory covers 316 km² and the inhabitants are 407.265 (2009 estimation). The population density is 1.368/km². The GDP per capita for 2009 is estimated at $ 19.111.

Constitutional Structure

The form of government is a parliamentary republic. It is divided into 86 elected local councils; administrative responsibility is distributed between the local councils and the central government.

25.2 Production of Raw Materials

Table 91: Production Data – Malta (World Mining Data, Weber and Zsak, 2008)

Resource		2002	2003	2004	2005	2006	Change 02/06	Change 05/06
Salt	(t)	13.020	13.020	6.000	6.000	6.000	−53,92	0.00

Table 92: Production of limestone – Malta (Metric tons) (USGS, 2008)

	2002	2003	2004	2005	2006
Limestone	1.200.000	1.200.000	1.200.000	1.200.000	1.200.000

25.3 Normative Basics

25.3.1 Primary Legal Basics

The primary legal basics of mineral extraction activity is the Malta Resources Authority Act XXV of 2000, as amended by Act XXIII of 2009. This act provides the establishment of an Authority of regulatory functions regarding resources relating to water, energy and mineral resources.

Besides that, the Land Acquisition (Public Purposes) Ordinance No. 1935 of 2004 is important. It regulates the acquisition of land for public purposes and establishes the procedures.

25.3.1.1 General Rules

Table 93: Structure of the Malta Resources Authority Act XXV of 2000

Part I Preliminary	Article 1 Short title	
	Article 2 Interpretation. *Amended by: XII. 2007.3;XXXII. 2007.16;XXIII. 2 009 101*	
Part II Establishment, Functions and Conduct of Affairs of the Authority	Article 3 Establishment and composition of the Malta Resources Authority	Sections 1–8
	Article 4 Functions of the Authority. *Amended by:XII. 2007.4*	Sections 1–3
	Article 5 Conduct of the affairs of the Authority	Sections 1–9
	Article 6 Relations between the Minister and the Authority	Sections 1–3
	Article 7 Legal personality and representation of the Authority	Sections 1–3
	Article 8 Provisions with respect to proceedings of the Authority	Sections 1–5
Part III Officers and Employees of the Authority	Article 9 Staff appointments	
	Article 10 Appointment and functions of officers and employees of the Authority	
	Article 11 Detailing of public officers for duty with the Authority	Sections 1–2
	Article 12 Status of public officers detailed for duty with the Authority	Sections 1–4
	Article 13 Offer of permanent employment with the Authority to public officers detailed for duty with the Authority	Sections 1–7

Part IV Financial Provisions	Article 14 Authority to meet expenditure out of revenue	Sections 1–5
	Article 15 Power to borrow or raise capital. *Amended by: L. No. 426 of 2007.*	Sections 1–2
	Article 16 Advances from Government	
	Article 17 Borrowing from Government	Sections 1–5
	Article 18 Estimates of the Authority	Sections 1–5
	Article 19 Expenditure to be according to approved estimates.	Sections 1–2
	Article 20 Publication of approved estimates.	
	Article 21 Accounts and audit.	Sections 1–4
	Article 22 Deposit of revenues and payments by the Authority	Sections 1–4
	Article 23 Contracts of supply of works. *Amended by: L. No. 426 of 2007*	
	Article 24 Annual report	
Part V Miscellaneous	Article 25 Appointment and functions of advisory committees	Sections 1–6
	Article 26 Licensing, etc., of activities. *Amended by: XII. 2007.5; L. No. 426 of 2007*	Sections 1–5
	Article 27 Persons deemed public officers	
	Article 28 Power to make regulations. *Amended by: XII. 2007.6*	Sections 1–2
	Article 29 Powers of service provider.	Sections 1–3
	Article 30 Enforcement powers of the Authority. *Complemented by Act XII. 2007.9. Amended by: L. No. 426 of 2007*	Sections 1–2
	Article 31 Imposition of administrative fines. *Complemented by Act XII. 2007.9. Amended by: II. 2009.58*	Sections 1–2
	Article 32 Power of Minister to make regulations in relation to criminal offences. *Complemented by Act.: XII. 2007.9. Amended by: L. No. 426 of 2007*	

	Article 33 Resources Appeals Board. *Amended by: XII. 2007.8*	Sections 1–5
	Article 34 Appeals. *Amended by: XII. 2007.10*	Sections 1–5
	Article 35 Powers and procedure of the Appeals Board. *Amended by: XII. 2007.11*	Sections 1–4
	Article 36 Appeal to the Court of Appeal. *Amended by: XII. 2007.12.*	
	Article 37 Savings. *Amended by: XII. 2007.13*	Sections 1–3
	Article 38 Exemption from liability. *Complemented by Act XII. 2007.14*	
	Article 39 Service providers. *Complemented by Act XXIII. 2 009 102* Cap. 500	Sections 1–9
First Schedule	(Article 5(2)) Directorates	Directorate for Energy Resources Regulation with responsibility for the regulation of all practices, relating to the generation, transmission, distribution, supply and use of energy, whatever the sources of any such energy.
		Directorate for Water Resources Regulation with responsibility for the regulation of all practices relating to water resources, drainage and sewage.
		Directorate for Minerals Resources Regulation with responsibility for the regulation of all practices relating to mineral resources.

According to Article 2 Resources Authority Act (RA) "mineral resources" includes any mineral, rock or sediment constituted of organic or inorganic compounds or substances extracted, mined or otherwise derived from the earth, including the seabed and the subsoil thereof.

Article 2 Land Acquisition Ordinance (LO) is dealing with "subsoil rights" (i.e. permits). Subsoil rights means the subjection of any land to the restrictive conditions regarding underground works and excavations (i.e. exploration, exploitation of minerals) referred to in Article 29. When for any public purpose any land is declared subject to subsoil rights, no owner can make any new or extend any existing underground work or excavation without the prior permission of the Board. When for any public purpose any land is declared to be subject to subsoil rights, no compensation is payable by reason only of the subjection of the land to such rights by the competent authority, of any underground work or excavation made.[1] If the competent authority opposes the grant of the permit, the Board shall determine the compensation payable by reason of the refusal of the competent authority to allow the erection of the proposed underground work or excavation. When compensation is paid under the provisions of this article, the Board must specify the area in respect of which the compensation is granted, and in such case no further compensation shall at any time thereafter become payable for subsoil rights (Article 29 LO).

25.3.1.2 Issuing of Permits

A person cannot carry out any activity/operation, or be engaged in an activity/operation, relating to mineral resources unless he has a licence (Article 26 (1) RA).

According to Article 4 (1) g) RA the competent authority is responsible in terms of investigation (e.g. exploration) relating to mineral resources.

According to Article 4 (1) b) RA the competent authority grants any permit, permit for the carrying out of operations relating to mineral resources. It must secure interconnectivity for the production of mineral resources (c).

---

[1] According Article 2 RO: "competent authority" means the Commissioner of Land; "Board" means the Board established under the provisions of this Ordinance.

Furthermore the authority must
- ensure fair competition in all such practices, operations and activities (d);
- establish minimum quality and security standards for any of the said practices, operations and activities and to regulate such measures as may be necessary to ensure public and private safety (e);
- secure and regulate the development and maintenance of efficient systems in order to satisfy, as economically as possible, all reasonable demands for the provision of the mineral resources (f).

According to Article 4 (2) RA the competent authority (c) in relation to mineral extraction must

- carry out such functions as may be authorised by the Minister in terms and for the purposes of the Continental Shelf Act;
- regulate all matters relating to the extraction of mineral resources;
- ensure the optimum utilisation of mineral resources and regulate the quality and quantity of minerals extracted.

The responsible Minister may after consultation with other authorities for the better carrying out of any of the provisions of this Act, make regulations to improve conditions concerning protection and extraction of mineral deposits. Besides that he may implement regulations to improve the services that may be required in relation to mining including time, manner, place and condition under which such services are to be provided (Article 28 (1) RA).

25.3.1.3 Authorities

The responsible Minister for mining will be represented by the Directorate for Minerals Resources Regulation. The Directorate is responsible for the regulation of all practices relating to mineral resources.

According to Article 4 (1) (a) RA the competent authority has to regulate, monitor and keep under review all practices, operations and activities relating to mineral resources.

25.3.1.4 Fees and Taxes

The price structure for the supply, storage or distribution of minerals resources is not regulated at present.[2]

---

2   http://www.mra.org.mt/minerals_tariffs.shtml

25.3.2 Additional Legal Basics

Table 94: List of Acts significant for raw materials – Malta

Name of Law	No. of Law and Year of Issuing
Environment Protection Act	Law No. 22 of 2009
Water services Corporation Act	ACT XXIII of 1991 as amended by Legal Notice 42 of 2010
Clean Air Act	Law No. 18 of 1967 as amended by Legal Notice 410 of 2007
Waste Management (Management of Waste from Extractive Industries and Backfilling) Regulations (part of Environment Protection Act	2009
Cultural Heritage Act	2002, last amended 2009
Public Administration Act	2009, last amended 194/2010
Public Registry Act	
Administration of Lands Act	2003, last amended 2006
Civil Code	1870, last amended 2009
Land Registration Act	1982, last amended 2007
Integrated Pollution Prevention & Control Regulations	
Quality of Water Intended for Human Consumption Regulations	2009
Strategic Environmental Assessment (Amendment) Regulations	2008
Building (Price Control) Act	
Development Planning Act	1992, last amended 2009

26 Moldavia

26.1 General Facts

The national territory covers 33 843 km² and the inhabitants are 3.567.500 (until 2009). The population density is 121,9/km². The GDP per capita for 2009 is estimated at $ 1.514.

Constitutional Structure

The form of government was before 1991: Soviet Socialist Republic of Moldova, since 1991: Republic of Moldova. Until 1978, the territory was divided into 40 raions and 4 cities, directly subordinated to the republican government – Chisinau, Balti, Bender and Tiraspol. Other four cities were added: Orhei, Ribnita, Soroca and Ungheni. Nowadays the total number of the cities is 21.

26.2 Production of Raw Materials

Table 95: Production Data – Moldavia (World Mining Data, Weber and Zsak, 2008)

Resources		2002	2003	2004	2005	2006	Change 02/06	Change 05/06
Gypsum	(t)	42.700	50.000	55.000	57.000	70.000	63,93	22,81

Table 96: Production of lime and gravel – Moldavia (Metric tons), (USGS, 2008)

	2002	2003	2004	2005	2006
Lime	3.300	2.880	1.911	1.900	1.900
Sand and gravel	700.000	1.000.000	1.600.000	1.900.000	2.100.000

26.3 Normative Basics

26.3.1 Primary Legal Basics

The primary legal basics of mineral extraction activity is the Mining Law ("Subsoil Code") No. 3 of 2009.

26.3.1.1 General Rules

Table 97: Structure of the Mining Law – Moldavia

Chapter I General Provisions	Article 1. Basics and definitions	
	Article 2. Purpose and scope of regulatory	
	Article 3. Law principles of the subsoil	
	Article 4. Legislation on subsoil	1–4
	Article 5. Participants in mining relations	1–3
	Article 6. The ownership of the subsoil minerals and mineral raw materials	1–3
	Article 7. State Fund of subsoil and State Fund of useful mineral deposits	
Chapter II State Regulation of Subsoil Use	Article 8. Central legal entities of public administration providing management of State in the field of subsoil use and protection	
	Article 9. Competence of Government in the subsoil use and protection	
	Article 10. Jurisdiction of the Ministry of Environment	
	Article 11. Jurisdiction Agency for Geology and Mineral Resources	1–2
	Article 12. Jurisdiction of local authorities in the subsoil use and protection	1–2

Chapter III Use of Subsoil	Article 13. The rights for use in the subsoil sector	1–3
	Article 14. The types of subsoil use	
	Article 15. Terms of subsoil use	1–5
	Article 16. Cause for issuing of right for use in the subsoil sectors	1–4
	Article 17. Antimonopoly requirements and ensuring publicity? and transparency in the use of subsoil	1–3
	Article 18. Allocation into sectors of subsoil use	1–8
	Article 19. Basic conditions of subsoil use included in the contract for the use in the subsoil sector	1–2
	Article 20. The competition for the right to use subsoil sectors	1–6
	Article 21. Allocation of subsoil sectors without contest	
	Article 22. Geological and mining perimeters	1–11
	Article 23. Assignment of subsoil sectors for geological research	1–6
	Article 24. Allocation of subsoil sectors for extracting minerals	1–9
	Article 25. Assignment of subsoil sectors for construction and/or operation of underground constructions not related to the extraction of minerals	1–2
	Article 26. Assignment of subsoil sectors for organization of protected geological targets	1–2
	Article 27. Assignment of sectors of subsoil to conduct mineralogy sampling, paleontology, and other geological materials sampling	
	Article 28. Assignment of the right to extract minerals and widely dispersed subsurface drinking water	1–3
	Article 29. Allocation of land to subsoil beneficiaries	
	Article 30. Assignment of the right for use in the subsoil	
	Article 31. The limitation, suspension or termination of the right for use in the subsoil sectors	1–3
	Article 32. The cause for limitation, suspension or termination of the right for use in the subsoil sectors	1–2
	Article 33. The termination of the right for use in the subsoil sectors	1–6
	Article 34. Liability for breach of contract clauses	1–2
	Article 35. Permit for extraction of minerals or natural mineral and drinking water	1–2
	Article 36. Temporary cessation of the permit	
	Article 37. Withdrawal of permit	

Chapter IV Rights and Obligations of Subsoil Beneficiaries	Article 38. Subsoil beneficiaries rights	
	Article 39. Obligations of subsoil beneficiaries	
	Article 40. Defending the rights of subsoil beneficiaries	1–2
	Article 41. Supporting local producers	1–2
	Article 42. State law on mineral raw materials purchase and its requisition	1–2
Chapter V Geological Subsoil Survey	Article 43. Geological subsoil research tasks	
	Article 44. Organization of the subsoil geology	1–3
	Article 45. Requirements for geological subsoil research	1–2
	Article 46. State registration and records of the geological subsoil research work	1–6
	Article 47. Types of geological subsoil research	1–5
	Article 48. Minerals and their types	1–3
	Article 49. Classification of reserves and resources of expected useful minerals	1–2
	Article 50. Geological information on subsoil	1–9
	Article 51. State fund on subsoil information	1–4
	Article 52. State Geological Survey	1–5
	Article 53. State cadastre of subsoil	1–4
	Article 54. State records of valuable mineral reserves	1–4
	Article 55. Discovery of valuable mineral deposits	1–2
Chapter VI Plan, Construction and Commissioning of the Business, Objectives and Structures related to the Use of Subsoil	Article 56. Characteristics of the planning enterprises, construction-related objectives and subsoil use	1–3
	Article 57. The main requirements for the planning, construction and commissioning and \ construction related to the use of subsoil	1–8
	Article 58. Normative losses when extracting useful minerals	1–4

Chapter VII Use of Exploitation of Subsoil Deposits of Useful Mineral Substances and Ancillary Purposes for Extraction of Minerals	Article 59. The exploitation of useful mineral deposits	1–3
	Article 60. Quotas for export of mineral raw materials and mineral extraction limits	
	Article 61. Requirements for the exploitation of deposits of useful minerals and primary processing of mineral raw materials	1–3
	Article 62. Characteristics of the exploitation of the hydrocarbons	1–6
	Article 63. Characteristics of the exploitation of the groundwater	1–8
	Article 64. Subsoil use not related to the extraction of minerals	1–4
	Article 65. Change to the loss of useful mineral reserves	
	Article 66. Mining Heritage	1–3
	Article 67. Liquidation and conservation of mineral excavation, underground construction targets not related to the extraction of minerals	1–7
	Article 68. Fund liquidation and recultivation	1–3
Chapter VIII Payment for the Use of Subsoil	Article 69. Types of payment for the use of subsoil	1–3
	Article 70. Regular payments for subsoil use	1–6
	Article 71. Settlement of geological exploration work made from the State Budget funds	1–4
Chapter IX Rational Use and Protection of Subsoil	Article 72. Basic requirements for rational use and subsoil protection	1–4
	Article 73. Conditions of building surfaces with deposits of useful minerals and other subsoil resources	1–4
	Article 74. State control over the rational use and protection of subsoil	1–2
	Article 75. State Mining Supervision	1–3
	Article 76. Geological Control of the State	1–2
Chapter X Safety at Work during Use of Subsoil	Article 77. The basic requirements for ensuring safety at work during use of subsoil	1–2
	Article 78. State supervision of insurance security of industrial use of subsoil	
Chapter XI Liability for Violations of the Subsoil and Settlements Disputes related to Use of Subsoil	Article 79. Liability for infringement of legislation on subsoil	1–2
	Article 80. Compensation for infringement	1–4
	Article 81. Subsoil use litigation matters	

Chapter XII International Cooperation in the Field of Subsoil Use and Protection	Article 82. The main directions and types of international cooperation considering the use and protection of subsoil	
Chapter XIII Transitional Provisions	Article 83. Action on the right to use subsoil sectors before coming into force of this Code	
Chapter XIV Final Provisions	Article 84. Entry into force of this Code	1–3
	Article 85. Bringing legislative and normative acts consistent with this Code	

Ownership of mineral rights

Minerals are the property of the state (Article 6 ML).

26.3.1.2 Issuing of Permits

According to Article 23 ML natural or legal persons who have explored a previously unknown deposits, which present valuable mineral resources, and persons who have discovered additional reserves of useful substances, are recognized as explorers.

According to Article 28 ML designing for the extraction of useful substances must be based on a proved mineral deposit. The following is required from the applicant.

- the extraction methods which exclude irrational exploitation of natural deposits, degradation of the explored deposits;
- security measures of life and health of workers and population,
- protection of subsoil and environment;
- a record keeping of extracted minerals;
- rational use of mineral resource;
- preparation of mining plans;
- a report on issues of reserves relevant for the State including quantitative and qualitative losses.

26.3.1.3 Authorities

According to Article 78 ML the competent authority responsible for supervising mining controls:

a) the requirements concerning secure works on subsoil use;
b) the conformity with rules of issued permits for the use and protection of subsoil;
c) the compliance with the requirement set out in the perimeters of mining projects;
d) the legal aspects of requirements related to subsoil users;
f) the rational use of underground deposits for other purposes;
h) the compliance with the established payment of subsoil use.

The competent authority responsible for supervising mining is entitled to stop mining and using underground projects in terms of insecure works or environment damage and to stop unauthorized use of subsoil

26.3.1.4 Fees and Taxes

Article 70 ML regulates the payment for use of subsoil.

26.3.2 Additional Legal Basics

Table 98: List of Acts significant for raw materials – Moldavia

Name of Law	No. of Law and Year of Issuing
Civil Code	Law No. 1107 of/ 2002 last amended by Law No. 598 of 2003
Land Code	Law No. 828-XII of 1991, last amended by Law No 3 of 2009
Forest Code	Law No. 887 of 1996, last amended by Law No. 327 of 2003
Water Code (Romanian)	Law No. 1532 of 1993, last amended in 2009
Housing Code	Law No. 2718-X of 1983, last amended in 2009
Law on Official Statistics	Law No. 412-XV of 2004
Law on Property	Law No.459-XII of 1991
Law on Licensing some types of Activities	Law No. 451-XV of 2001

Name of Law	No. of Law and Year of Issuing
Law of Business and Enterprises	Law N.845-XII of 1992, last amended in 1997
Law on State Registration of Enterprises and Organizations	Law No. 1265-XIV of 05.10.2000
Law on Customs Tariff	Law No. 138-XIII of 1997
Law on Waste Production and Waste	Law No. 1347 of 1997
Law on Environmental Protection	Law No. 1515-XII of 1993, last amended by Law No. 59 of 2003
Law on Areas Protection for River Waters	Law No. 440 of 1995
Law on Ecological Examination and Environmental Impact Assessment	Law No. 851 of 1996, last amended by Law No. 59 of 2003
Law on Natural Resources	Law No. 1102 of 1997
Law on Harmful Products and Substances	Law No. 1236 of 1997
Law on Atmospherical Air Protection	Law No. 1422 of 1997
Law on Payment for Environmental Pollution	Law No. 1540 of 1998, last amended by Law No. 1556 of 2002
Law on Protected Natural Areas of State	Law No. 1538 of 25.02.1998, last amended in 2009
Law on Drinking Water	Law No. 272 of 1999
Law on Access at Information	Law No. 982 of 2000

27 Netherlands

27.1 General Facts

The national territory covers 41.526 km² and the inhabitants are 16.500.156 (2009 estimation). The population density is 318/km². The GDP per capita for 2009 is estimated at $ 48.223.

Constitutional Structure

The form of government is a parliament democracy and constitutional monarchy. The Netherlands are divided into 12 administrative regions (provinces), each under a governor. All provinces are subdivided into 430 municipalities. The country is also subdivided in water districts which are governed by a water board.

27.2 Production of Raw Materials

Table 99: Production Data – Netherlands (World Mining Data, Weber and Zsak, 2008)

Resources		2002	2003	2004	2005	2006	Change 02/06	Change 05/06
Aluminium	(t)	284.400	282.800	327.000	333.800	294.000	3,38	−11,92
Cadmium	(t)	485	495	572	570	560	15,46	−1,75
Salt	(t)	5.773.036	5.980.114	5.895.648	6.155.651	6.156.000	6,63	0,01
Nat. Gas	(Mm³)	70.715	68.765	77.544	73.116	73.000	3,23	−0,16
Oil	(t)	3.323.544	3.147.468	3.150.972	2.137.440	2.137.000	−35,70	−0,02

Table 100: Aggregates Production – Annual Statistics/Netherlands, 22 April 2008 (UEPG 2008)

Sand & Gravel (1)		Crushed rock (2)		Marine Aggregates		Recycled Aggregates (3)		Manufactured Aggregates (4)	
2005	2006	2005	2006	2005	2006	2005	2006	2005	2006
94,2	44,5	n.a.	n.a.	n.a.	50,0	25,0	25,0	n.a.	n.a.

(1) Sand and Gravel: sold production including crushed gravel
(2) Crushed rock: sold production (excluding crushed gravel)
(3) Recycled Aggregates: materials coming from construction and demolition waste used in aggregates market
(4) Manufactured aggregates include blast-furnace-slag, electric-arc-furnace-slag, incinerator bottom ash (IBA), pulverised fuel ash (PFA)

27.3 Normative Basics

27.3.1 Primary Legal Basics

The primary legal basics of mineral extraction activity is the Mining Law ("Mining Act") 2003 and the Excavation Act 2002. The latter is regulating the surface minerals, e. g. industrial and construction minerals. The Excavation Act has been restricted since 2008 and in 2011/2012 the Extraction Act will be abolished. The mineral permit will be integrated in the new Environmental Protection Act (land based excavations) and the Water Act (excavations in State Waters).

27.3.1.1 General Rules

Table 101: Structure of the Mining Law (2003)

Chapter 1 Definitions and General Provisions		Article 1	a–p
		Article 2	1–3
		Article 3	1–4
		Article 4	
		Article 5	
Chapter 2 Permits for Exploration and Production	2.1 General Rules	Article 6	1–3
		Article 7	1–3
		Article 8	
		Article 9	1–4
		Article 10	1–3
	2.2. Restrictions and Conditions	Article 11	1–5
		Article 12	
		Article 13	1–3
	2.3. Procedure	Article 14	
		Article 15	1–5
		Article 16	
		Article 17	1–3
	2.4. Amendment, Transfer and Withdrawal	Article 18	1–4
		Article 19	a–b
		Article 20	1–3
		Article 21	1–6

	2.5. Special Provisions	Article 22	1–10
		Article 23	1–2
		Article 24	
Chapter 3 Permits for the Storage of Substances		Article 25	1–2
		Article 26	1–2
		Article 27	1–3
		Article 28	a–b
		Article 29	1–2
		Article 30	
		Article 31	
		Article 32	
Chapter 4 Ensuring that the Activities are executed properly	4.1. General Obligations	Article 33	a–d
		Article 34	1–5
		Article 35	1–3
		Article 36	1–3
		Article 37	1–2
		Article 38	
		Article 39	a–b
		Article 40	1–9
		Article 41	1–4
		Article 42	1–2
		Article 43	1–2
		Article 44	1–5
		Article 45	1–2
	4.2. Financial Security	Article 46	1–4
		Article 47	1–2
		Article 48	1–2
	4.3. Further Rules	Article 49	1–6
		Article 50	
		Article 51	1–3
		Article 52	1–3
Chapter 5 Financial Conditions	5.1.1.1. General	Article 53	
		Article 54	a–f
		Article 55	1–4
	5.1.1.2. Surface Rentals	Article 56	1–2
		Article 57	1–2

	Article 58	1–3
	Article 59	1–2
5.1.1.3 Tax	Article 60	1–4
	Article 61	
	Article 62	1–4
	Article 63	1–5
	Article 64	1–2
5.1.1.4. Profit Share	Article 65	
	Article 66	1–3
	Article 67	1–4
	Article 68	1–4
	Article 69	1–4
	Article 70	1–2
5.1.1.5. Assessment and Collection	Article 71	
	Article 72	
	Article 73	1–2
	Article 74	
5.1.2. Payments to the Province	Article 75	
	Article 76	1–2
	Article 77	1–2
	Article 78	
	Article 79	
	Article 80	
5.2.1. State Participation in Exploration Licences for Hydrocarbons for the Seawardside	Article 81	a–b
	Article 82	
	Article 83	1–3
	Article 84	a–d
	Article 85	a–c
	Article 86	1–2
	Article 87	a–c
	Article 88	a–b
5.2.2. State Participation in Production Licences for Hydrocarbons	Article 89	a–c
	Article 90	1–2
	Article 91	1–3
	Article 92	1–3
	Article 93	1–2

		Article 94	1–3
		Article 95	1–2
		Article 96	
		Article 97	a–b
	5.3. Payments in Connection with Licences other than for Exploration or Production and State Participation in the Storage of Substances	Article 98	1–4
		Article 99	
		Article 100	1–2
		Article 101	1–4
	5.4. The Provision of Security	Article 102	1–3
	5.5. Implementation Rules	Article 103	1–2
	5.6. Scientific Research	Article 104	
Chapter 6 Advisors	6.1. The Mining Council	Article 105	1–3
		Article 106	1–3
		Article 107	1–2
		Article 108	
		Article 109	1–2
		Article 110	
		Article 111	
		Article 112	
	6.2. The Soil Movement Technical Committee	Article 113	a–c
		Article 114	1–3
		Article 115	1–3
		Article 116	1–4
		Article 117	1–5
		Article 118	1–2
		Article 119	1–5
		Article 120	1–3
		Article 121	
		Article 122	
Chapter 7 Reporting		Article 123	1–6
Chapter 8 Supervision and Enforcement	8.1. State Supervision and Enforcement	Article 124	1–2
		Article 125	1–2
		Article 126	1–2
		Article 127	

		Article 128	1–2
		Article 129	1–2
		Article 130	
	8.2. Supervision in specific Cases	Article 131	1–3
		Article 132	
		Article 133	1–2
Chapter 9 Guarantee Fund Mining Damage	9.1. General Provisions	Article 134	1–2
		Article 135	1–7
		Article 136	1–2
	9.2. Compensation for Damages in Cases of Insolvency	Article 137	a–c
		Article 138	1–2
		Article 139	1–2
	9.3. Preliminary Payments	Article 140	1–2
		Article 141	1–2
Chapter 10 Legal Protection		Article 142	1–3
Chapter 11 Transitional Provisions		Article 143	1–7
		Article 144	
		Article 145	1–3
		Article 146	1–4
		Article 147	1–4
		Article 148	1–2
		Article 149	1–2
		Article 150	1–3
		Article 151	
		Article 152	1–3
		Article 153	1–4
		Article 154	
		Article 155	1–2
		Article 156	
		Article 157	
		Article 158	1–4
		Article 159	a–c
		Article 160	1–2
		Article 161	
		Article 162	1–2

		Article 163	
		Article 164	1–3
		Article 165	
		Article 166	
		Article 167	1–3
Chapter 12 Withdrawal and Amendment of certain Acts	12.1. Ministry of Economic Affairs	Article 168	a–n
		Article 169	
		Article 170	
		Article 171	
	12.2. Ministry of Transport and Public Works	Article 172	
		Article 173	a–d
		Article 174	1–2
	12.3. Ministry of Justice	Article 175	
		Article 176	
		Article 177	a–b
		Article 177 amended	1–5
		Article 178	
		Article 179	a–c
	12.4. Ministry of Housing, Planning and the Environment	Article 180	
		Article 181	
		Article 182	1–2
		Article 183	a–f
	12.5. Ministry of Social Services and Employment	Article 184	a–b
		Article 185	
		Article 186	a–b
		Article 187	1–2
		Article 188	a–b
Chapter 13 Final Provisions		Article 189	
		Article 190	
		Article 191	
		Article 192	
		Article 193	

According to Article 1 ML minerals are substances of organic origin, present in the subsoil, in a concentration or deposit which is there by natural origin, in solid, liquid or gaseous form, with the exception of limestone, gravel, sand, clay.

Ownership of mineral rights

Minerals are the property of the state Article 3 (1) ML. The ownership of minerals produced under a production permit is transferred to the permit-holder (by the production of the said minerals). Similarly applies to minerals that are produced from the subsoil (in the form of samples) under the terms of an exploration permit (2). In all negotiations connected with the ownership of minerals, the state is represented by the competent authority (4). According to Article 4 the rightful claimant to the surface of the earth is obliged to permit the holder of a permit for the exploration for or the production of minerals to explore for or produce minerals, without prejudice to the right of the rightful claimant to the surface to receive compensation for any damage caused by these activities.

Excavation Act

The Dutch state is the owner of the shells, gravel, sand and clay at or near the surface of the Continental Shelf (Article 4b Excavation Act). The Dutch state is also the owner of the seabed within its borders and the large inland waters (rivers). For land-based extraction, the operator needs an extraction permit, and the operator must have the permit of the landowner. It is also possible for the operator to purchase land. Extraction companies do not have to own the land as long as they have the permit of the landowner.[1]

27.3.1.2 Issuing of Permits

Exploration

Without permit from the responsible minister, it is prohibited to explore for minerals and to produce minerals (Article 6 ML). A production permit will only be granted if it is feasible that the minerals within the permitted area, are economically producible (Article 8 ML). Without prejudice to Articles 7 and 8a permit can only be refused: a) on the basis of the technical or financial capability of the applicant, b) on the basis of the manner in which the applicant intends to carry out the activities for which the permit is applied for c) on the basis of lack of efficiency and responsibility, which shall include sense of responsibility for society (Article 9 ML).

---

1 Ike, P. (2004), Country Report Netherlands, in: Department of Mining and Tunnelling, lc.

Excavation Act

According to the Excavation Act 2002 it is not necessary to have a permit for the exploration of surface minerals (opencast mining). No payments must be made to either the government or the landowner. The operator usually asks for permit of the landowner.

Exploitation

The holder of an exploration permit, who has demonstrated the presence of the relevant minerals, having submitted an application during the period of the validity of the exploration permit, may be granted a production permit for those minerals in the permitted area (Article 10 ML). If it applies to an area that is not delimited according to Article 11 (5) ML, and the presence of the relevant minerals has been shown only in a part of the area, the production permit shall only be granted for that part of the area for which granting is justified from a geological point of view.

According to Article 11 ML:

1. A permit specifies for which of the activities referred to in Article 6 and for which minerals it is valid. If in a production permit it is stated, that it applies to certain minerals, it will also apply to other minerals that are inevitably produced in conjunction with those certain minerals.
2. A permit specifies for what period it is valid. This is done such that the period is no longer than necessary for carrying out the activities for which it is granted.
3. A permit specifies for which area it is applicable. For the delineation of this area, the limits on the surface are indicated. Unless the permit specifies otherwise, the area shall consist of the surface area indicated and its subsoil.
4. The delineation of the area must be done in such a manner that the activities can be carried out in the optimum possible manner from a technical and economical point of view.

According to Article 34 (1) ML the production of minerals must be carried out according to a production plan. The holder of a production permit submits a production plan to the responsible Minister. The production plan needs an approval. The production plan sets forth in respect of each deposit within the permit area at least a description of (Article 35 (1) ML):

a. the anticipated volume of minerals present and the location thereof;
b. the commencement and duration of the production;
c. the manner of production and the activities relating thereto;

d. the volume of minerals to be produced annually;
e. the cost on an annual basis of the mineral production;

Excavation Act[2]

Article 10 of the Excavation Act states that: "at the preparation of a decision concerning granting or refusing, adjustment or revokement of a permission, paragraph 3.5.2 up to and including 3.5.5 and 3.5.6 of the Administrative Law Act is applicable". In fact, the Administrative Law Act regulates the procedure for the excavation permit.

Extraction companies must apply for an extraction permit with the province or a regional directorate of the Directorate General for Public Works and Water Management (State Waters). An extraction permit is provided when the request is in accordance with the provincial Regional Spatial Plan (Spatial Planning Act) and/or Regional Mineral Extraction Plan (provinces and/or state waters).

The provincial executive plays a central role in the decision process. The Excavation Law is added to Article 13 of the Environmental Protection Act. Through this, the coordination and regulation of chapter 14 of the Environmental Protection Act is applicable in case more must be granted for the same excavation. By request of the applicant, the authorized body needs to be conducive to coordinated consideration of the application. Through the Environmental Protection Act an environmental permit is needed for excavations. Under this Environmental Protection Act several laws are categorized, partly with respect to content, partly procedural (Air Pollution Act, Waste Substances Act, Nuisance Act, etc.).

Restoration

The conditions that can be attached to excavation permits are defined in the Excavation Act and in the provincial ordinances. The Excavation Act has a decentralised set up (framework act) and is worked out in 12 provincial ordinances – which differ in details – and in one national ordinance for State Waters.

27.3.1.3 Authorities

The main responsible authority for mining is the Minister of Economic Affairs.

---

2   Ibidem.

According to Article 126 ML State supervision of mines is responsible to the inspector-general of mines. The inspector-general of mines issues an annual report before 1st May to the competent Minister on the operations of the state supervision of mines during the past year.

Excavation Act

Enforcement is regulated in the Excavation Act (article 21 g – 25) and the Environmental Protection Act (article 18.4 to 18.12 and article 18.14 and 18.16). Public servants of the permit granting government take care of the enforcement practices.

27.3.1.4 Fees and Taxes

Excavation Act

Since 1998 the provinces have been authorised to levy tax up to € 0.10/m³ to meet the costs of compensation measures in case of far-reaching excavations. The application regulations are restricted to restoration and aftercare of the abandoned working site and restoration of the surrounding areas. If compensation has to be paid elsewhere, a single payment can be drawn from the proceeds of the levy.³

27.3.2 Additional Legal Basics

Table 102: List of Acts significant for raw materials – Netherlands

Name of Law	No. of the Law and Year of Issuing
The Environmental Protection Act	1994
Spatial Planning Act	
Water Act	
Air Pollution Act	
Noise Abatement Act	
Building Material Decree	
Soil Protection Act	1996

---

3   Ibidem.

28 Norway

28.1 General Facts

The national territory covers 323,758 km², excluding Svalbard and Jan Mayen, and the inhabitants are 4,7 million. The population density is 12/km². In 2009 the GDP per capita amounted to $ 79.085.

Constitutional Structure

The form of government is a constitutional monarchy. The executive power is entrusted to the government; the Parliament (Stortinget) is responsible. Norway is divided into a national level, regional level (five regions), provincial level (19 provinces), district and municipal level (433 municipalities).

28.2 Production of Raw Materials

Table 103: Production Data – Norway (World Mining Data, Weber and Zsak, 2008)

Resources		2002	2003	2004	2005	2006	Change 02/06	Change 05/06
Iron	(t)	476.000	398.868	573.000	448.000	396.800	−16,64	−11,43
Cobalt	(t)	100	0	0	0	0	−100,00	
Nickel	(t)	1.540	200	200	100	200	−87,01	100,00
Titanium	(t)	367.019	377.968	387.500	310.000	374.000	1,90	20,65
Aluminium	(t)	1.044.000	1.192.000	1.321.700	1.376.000	1.381.000	32,28	0,36
Cadmium	(t)	215	323	141	153	125	−41,86	−18,30
Silver	(kg)	1.200	1.000	1.000	900	700	−41,67	−22,22
Feldspar	(t)	80.000	70.000	100.000	210.000	65.000	−18,75	−69,05
Graphite	(t)	8.000	1.000	6.000	7.000	9.000	12,50	28,57
Sulphur	(t)	38.000	30.000	20.000	20.000	18.000	−52,63	−10,00
Talc	(t)	27.000	28.000	28.000	27.000	26.000	−3,70	−3,70
Steam Coal	(t)	2.200.000	2.900.000	2.904.000	1.440.000	2.400.000	9,09	66,67

Resources		2002	2003	2004	2005	2006	Change 02/06	Change 05/06
Natural Gas	(Mm³)	68.300	76.900	80.600	88.600	91.500	33,97	3,27
Oil	(t)	159.200.000	151.700.000	146.800.000	137.700.000	130.000.000	−18,34	−5,59

Table 104: Aggregates Production – Annual Statistics/Norway, 22 April 2008, quantities in million tonnes (UEPG 2008)

Sand & Gravel (1)		Crushed rock (2)		Marine Aggregates		Recycled Aggregates (3)		Manufactured Aggregates (4)	
2005	2006	2005	2006	2005	2006	2005	2006	2005	2006
15,0	13,4	38,0	45,0	n.a.	0,0	0,2	n.a.	n.a.	n.a.

(1) Sand and Gravel: sold production including crushed gravel
(2) Crushed rock: sold production (excluding crushed gravel
(3) Recycled Aggregates: materials coming from construction and demolition waste used in aggregates market
(4) Manufactured aggregates include blast-furnace-slag, electric-arc-furnace-slag, incinerator bottom ash (IBA), pulverised fuel ash (PFA)

28.3 Normative Basics

28.3.1 Primary Legal Basics

The primary legal basics of mineral extraction activity is the Mining Act No. 101 of 2009 (relating to the acquisition and extraction of mineral resources).

Table 105: Structure of the Mining Law – Norway

Chapter 1 Introductory Provisions	Article 1	Purpose of the Act
	Article 2	Considerations relating to the aAdministration and use of mirResources
	Article 3	Substantive scope
	Article 4	Geographical scope
	Article 5	Relationship with other legislation
	Article 6	Relationship with international legislation
	Article 7	Categories of minerals used in the Act
Chapter 2 Searching	Article 8	Right to search
	Article 9	Scope of the right to search
	Article 10	Duty to give notice

Chapter 3 **Exploration of Minerals owned by a Landowner**	Article 11	Agreement granting an exploration permit
	Article 12	Pilot extraction
Chapter 4 **Exploration of Minerals Owned by the State**	Article 13	Application for an exploration permit
	Article 14	Relationship with other right holders
	Article 15	Exploration Area
	Article 16	Priority
	Article 17	Applications relating to exploration in the Finnmark county
	Article 18	Duty to give notice
	Article 19	Content of the exploration permit
	Article 20	Pilot extraction
	Article 21	Provision of security
	Article 22	Expiry of an exploration permit
	Article 23	Extension of an Exploration Permit
	Article 24	Quarantine period in connection with the expiry of an exploration permit or an extraction permit
	Article 25	Exploration reports, measurement data and sample materials
	Article 26	Transfer of an exploration permit
	Article 27	Identification of persons and companies with an applicant
Chapter 5 **Extraction Permits for Minerals Owned by a Landowner**	Article 28	Agreement granting an extraction permit
Chapter 6 **Extraction Permits for Minerals Owned by the State**	Article 29	Application for an extraction permit
	Article 30	Extraction permits relating to Finnmark
	Article 31	Extraction area
	Article 32	Content of an extraction permit
	Article 33	Expiry of an extraction permit
	Article 34	Extension of an extraction permit
	Article 35	Transfer of an extraction permit
	Article 36	Registration
Chapter 7 **Compulsory Acquisition**	Article 37	Compulsory acquisition of minerals owned by a land owner
	Article 38	Compulsory acquisition of land and rights for the exploration and extraction of minerals owned by the State
	Article 39	Compensation
	Article 40	General rules on compulsory acquisition

Chapter 8 Operations	Article 41	Good mining practice
	Article 42	Extraction subject to notification
	Article 43	Extraction subject to licensing (operating license)
	Article 44	Notification of commencement and suspension of operations
	Article 45	Expiry of an operating permit
	Article 46	Information on operations
Chapter 9 General Provisions	Article 47	Areas exempted from searching and exploration
	Article 48	Duty to exercise caution
	Article 49	Duty to implement safety measures
	Article 50	Duty to clean up
	Article 51	Coverage of clean-up costs and safety-measure costs
	Article 52	Duty to compensate
	Article 53	Valuation proceedings
	Article 54	Requirements relating to applications and the use of electronic means of communication
Chapter 10 Charges and Fees	Article 55	Processing and supervision charges
	Article 56	Annual fees paid to the State
	Article 57	Annual land owner fee
	Article 58	Increased land owner fee in Finnmark
Chapter 11 Supervision	Article 59	Supervision
	Article 60	Inspection and the rRight to information in connection with supervision
	Article 61	Internal control
Chapter 12 Administrative Measures and Administrative Sanctions	Article 62	Orders
	Article 63	Immediate implementation by the directorate of mining
	Article 64	Temporary suspension of operations
	Article 65	Amendment and revocation of permits, etc.
	Article 66	Enforcement penalty
	Article 67	Infringement penalty
Chapter 13 Entry into Force and Transitional Provisions	Article 68	Entry into force
	Article 69	Transitional provisions: items 1–8
	Article 70	Amendments to other acts

Ownership of mineral rights

According to Article 7, "minerals owned by the State" include:

a) the metals with a specific gravity of 5 g/cm³ or greater, including chromium, manganese, molybdenum, niobium, vanadium, iron, nickel, copper, zinc, silver, gold, cobalt, lead, platinum, tin, zinc, zirconium, tungsten, uranium, cadmium and thorium, and their ores. Alluvial gold, however, does not fall within the definition;
b) the metals titanium and arsenic, and their ores;
c) the minerals pyrrhotite and pyrite.

The term "minerals owned by a land owner" include all minerals that are not owned by the State (i.e. industrial minerals and construction minerals).

28.3.1.1 Issuing of Permits

Prospection and Exploration

Any party has the right to search (i.e. prospect) for mineral deposits on another party's land, subject to the limitations set out in this Act and other legislation (Article 8 ML). The searching party can undertake such works on the surface of the land as are necessary to establish the existence of mineral deposits (Article 9 ML).

The searching party must give notice to the land owner and the user of the land one week before a search is begun, at the latest (Article 10 M).

Exploration of minerals owned by a land owner – Agreement granting an exploration permit (Article 12 ML)

Any party wishing to explore deposits of minerals owned by a land owner must enter into an agreement with the land owner (Article 11 ML). If no agreement is reached, an application may be made for compulsory acquisition, (see Article 37 ML). In addition to an agreement with the land owner, pilot extraction requires a special permit from the Directorate of Mining. The term "Pilot extraction" refers extraction that is necessary to assess the commercial viability of a deposit. Except in special cases, a permit for pilot extraction may not be granted for the extraction of more than 2.000 m³ of matter (Article 20 ML).

Exploration of minerals owned by the State – Application for an exploration permit (Article 13 ML)

A party that wishes to secure a right to explore deposits of minerals owned by the State must apply to the Directorate of Mining for an exploration permit. An exploration permit can be refused if the applicant has previously breached material provisions imposed by or pursuant to this Act. The exploring party may be granted only one exploration permit for a given area. A party that holds an extraction permit for deposits of minerals owned by the State cannot apply for an exploration permit for any part of the same area.

The Ministry can issue regulations concerning the exploration area, including its shape and size (Article 15 ML). The priority of the exploring party in respect of an exploration area should be calculated from the day on which the application for an exploration permit is received by the Directorate of Mining (Article 16 ML). An exploring party must give written notice of exploration to the Directorate of Mining, the land owner and the user of the land at least three weeks before the work is begun. The Directorate of Mining gives notice to the municipality, the county municipality and the county governor. The notice contains a plan for the work to be carried out and for the access to and within the exploration area (Article 18 ML).

An exploring party may undertake the exploration needed to assess whether there is a deposit of minerals of such an abundance, size and nature that it may be assumed commercially viable, or to become commercially viable within a reasonable period of time. The exploration permit grants such access to the land (including temporary storage space) as it is necessary to undertake the exploration. The exploration permit does not grant a right of way. Measures in the ground that may cause considerable damage may only be implemented with the consent of the land owner and the user of the land (Article 19 ML).

Expiry, extension and transfer of an exploration permit

An exploration permit expires 7 years after the date on which the permit was issued, or upon the expiry of an extended deadline pursuant to section 23. The 7-year period begins to run on the date on which the exploring party obtains best priority in respect of the exploration area (Article 22 ML). The Directorate of Mining may extend the deadline by up to 3 years if the applicant substantiates that exploration cannot be completed before the deadline due to extraordinary circumstances which are not the fault of the applicant (Article 23 ML). An exploration permit may be transferred. The transfer requires the approval of the Directorate of Mining. Articles 13 and 27 ML apply correspondingly in the event of a transfer. An application for approval must be sent to the Directorate of Mining (Article 26 ML).

Exploitation permit

Extraction permits for minerals owned by a land owner – Agreement granting an extraction permit (Article 28 ML). Any party wishing to extract deposits of minerals owned by a land owner must enter into an agreement with the land owner. If no agreement is reached, an application may be made for compulsory acquisition (see Article 37 ML).

Extraction permits for minerals owned by the State – Application for an extraction permit

The exploring party with best priority may apply to the Directorate of Mining for an extraction permit. An extraction permit is granted if the applicant substantiates that the exploration area contains a deposit of minerals owned by the State which is of such an abundance, size and nature that it may be assumed to be commercially viable, or to become commercially viable within a reasonable period of time. Multiple extraction permits may not be granted for the same area (Article 29 ML). If the application relates to areas that are not covered by the area of the exploration permit, the application may also count as an application for an exploration permit. The application has priority from the date on which it was received by the Directorate of Mining. Exploring parties with identical priority have an equal right to apply for an extraction permit. If extraction permits are granted to more than one party, the parties hold the permit jointly, unless they agree on a different arrangement.

The Ministry can issue regulations concerning the extraction area, including its shape, size and marking (Article 31 ML). The extracting party may extract and utilise all deposits of minerals owned by the State in the extraction area. Deposits of minerals owned by a land owner may be extracted to the extent that this is necessary to extract deposits of minerals owned by the State (Article 32 ML).

Expiry, transfer of an extraction permit

An extraction permit may expire if: a) an operating permit pursuant to section 43 is not granted within 10 years of the date on which the extraction permit was granted; b) the extraction does not require an operating permit pursuant to the Act, and more than 10 years have passed since the date on which the extraction permit was granted without operations being initiated (Article

33 ML). Article 26 ML applies correspondingly to a transfer of an extraction permit (Article 35 ML).

Registration

An extraction permit must be registered in the Land Register. The transfers and the charges of extraction permits are afforded legal protection when registered in the Land Register. The same applies to other legal rights created over an extraction permit (Article 36 ML).

Compulsory acquisition of minerals owned by a land owner

Any party may apply to the Directorate of Mining for a permit to acquire compulsorily the land and rights needed to explore whether there is a deposit of minerals owned by a land owner that is of such an abundance, size and nature that the deposit may be assumed to be commercially viable, or to become commercially viable within a reasonable period of time (Article 37 ML). Any party may apply to the Ministry for a permit to acquire compulsorily:

a) a deposit of minerals owned by a land owner;
b) the land and rights needed for extraction, including access to the deposit;
c) the land and rights needed for processing of minerals owned by a land owner.

In the assessment of whether or not compulsory acquisition should be granted, emphasis must be given to whether the applicant has explored the deposit.

An exploring party may apply to the Directorate of Mining for a permit to acquire compulsorily the land and rights needed to be able to undertake exploration (Article 38 ML). If another exploring party has better priority in respect of the exploration area, the compulsory-acquisition permit may only be granted if that party gives its consent. A party that is extracting a deposit of minerals owned by the State may apply to the Ministry for a permit to acquire compulsorily:

a) the land and rights needed for extraction; and
b) the land and rights needed for the processing of minerals.

Compensation (Article 39 ML)

In the event of the compulsory acquisition of a right to extract a deposit of minerals owned by a land owner pursuant to section 37, second paragraph, sub-paragraph a) a compensation must be set on the basis of the market price and without regard to the foreseeability requirement in sections 5 and 6 of the Act of 6 April 1984 No. 17 relating to compensation in connection with the compulsory acquisition of real property (the Compulsory Acquisition Compensation Act). The compensation must take the form of a charge per extracted unit of the mineral, unless special reasons indicate that a different solution is appropriate. A minimum charge may be set that is to be paid regardless of the production volume. Awarded compensation is set as annual payment. However, a one-off compensation payment may be set if there are special reasons for doing so.

The provisions of the Act of 23 October 1959 No. 3 relating to the compulsory acquisition of real property (the Compulsory Acquisition Act), and the Compulsory Acquisition Compensation Act apply insofar as they are relevant to cases of compulsory acquisition (Article 40 ML). The Directorate of Mining must be notified for the extraction of more than 500 m^3 of matter. The notice must be sent at least 30 days before works are begun. In special cases, the Directorate of Mining may require the submission of a plan of operations. The Directorate of Mining can decide that operations should not begin until the plan of operations has been approved (Article 42 ML).

Extraction subject to permit (operating permit)

According to Article 43 ML extraction of mineral deposits totalling more than 10.000 m^3 requires an operating permit from the Directorate of Mining. An operating permit can only be granted to a party that holds an extraction permit. In the assessment of whether an operating permit should be granted, emphasis must be given to whether the applicant is qualified to extract the deposit. The area of operation must be fixed in the permit. An application for an operating permit must contain a plan of operations. A permit may be limited in time and it may be made subject to review after a specified period of time. The permit may in any event be reviewed every 10th year. An operating permit expires if operations have not begun, at the latest, within 5 years of the operating permit being granted. The same applies if operations are discontinued for more than one year. The Directorate of Mining may extend the deadlines in the first and second sentences. The deadline in the second sentence may be extended by up to 3 years at a time (Article 45 ML).

Areas exempted from searching and exploration (Article 47 ML)

Prospecting and exploration is prohibited in areas covered by the Act of 5 June 2009 No. 35 relating to natural amenities in Oslo and surrounding municipalities (the Natural Amenities Act). A prospecting or an exploring party should not, without the consent of the land owner, the user of the land and the relevant authority, prospect or explore the following:

a) cultivated land;
b) industrial areas, including soil extraction sites, quarries and mines in operation;
c) areas lying less than 100 m from buildings used as permanent or temporary residences, including holiday cabins;
d) areas belonging to facilities that are of public utility, and locations lying less than 20 m from such facilities;
e) areas belonging to military facilities or used for military exercises;
f) abandoned mining areas, including waste rock tips and tailing dams or landfills.

The exploring party, the extracting party and the working party in respect of a mineral deposit must ensure that the area is properly cleaned up, both while operations are in progress and after they have been completed. The Directorate of Mining can set a deadline for the completion of clean-up works (Article 50 ML). The Directorate of Mining may order a party that wishes to undertake, or has initiated, exploration (including sample extraction), or operations on mineral deposits to provide financial security for the implementation of safety measures pursuant to section 49 and clean-up measures pursuant to section 50 (Article 51 ML).

28.3.1.2 Authorities

The responsible authority is the Directorate of Mining and the responsible ministry is the Ministry of Trade and Industry.

The Directorate of Mining supervises the following matters (Article 59 ML):

a) that exploration, including pilot extraction, and operations are undertaken in accordance with good mining practice pursuant to the requirements of the Act;
b) that conditions, approved plans of operations and orders given in or pursuant to this Act are complied with;
c) that the duty to implement safety measures and the duty to clean up are complied with; and
d) that operations do not result in unnecessary pollution or unnecessary damage to the environment.

The Directorate of Mining must, while conducting inspections, be given unhindered access to areas and facilities, and be allowed to conduct necessary examination of such areas and facilities (Article 60 ML).

28.3.1.3 Fees and Taxes

According to Article 55 ML the Ministry may issue regulations concerning charges for the processing of applications and for supervision pursuant to this Act.

Annual fees paid to the State (Article 56 ML)

Parties that are exploring or extracting deposits of minerals owned by the State must pay an annual fee to the State for their exploration and extraction permits. The fees must be paid in advance to the Directorate of Mining, by 15 January. If the fees are not paid in time, an additional fee of 50 % must be paid by 30 April of the same year. The Directorate of Mining may in special cases make exceptions from this provision. The permit lapses if the additional fee is not paid before the deadline.

Annual land owner fee (Article 57ML)

A party that is extracting a deposit of minerals owned by the State must pay the land owner an annual fee of 0.5 % of the sales value of that which is extracted. The fee for each year falls due for payment on 31 March of the following year. If there are several land owners in the extraction area, the fee will be divided among them in proportion to the land owned by each of them in the extraction. In the Finmark county there is an additional fee to the Finmark property of 0,25% of the sales value of that which is extracted.

28.3.2 Additional Legal Basics

Table 106: List of Acts significant for Raw Materials – Norway

Name of Law	No. of Law and Year of Issuing
Pollution Control Act	Law No. 6 of 1981, as amended by Law 2003
Nature Diversity Act	Law No. 100 of 2009
Property Concession Act	Law No. 98 of 2003
Land Expropriation Act	No. 1 of 1952
Forestry and Forestry Protection Act	Law No. 33 of 1965, as amended by Law No. 55 of 2005

28.3 Normative Basics

Name of Law	No. of Law and Year of Issuing
Cultural Monuments Act	Law No. 50 of 1978, as amended by Law No. 96 of 1992
Public Administration Act	1967, last amended 86 of 2003
Competition Act	2004
Act relating to value Added tax	1969
Act relating to protection against pollution and relating to waste	Law No. 6 of 1981, last amended by Law No. 83 of 1999
Environmental Information Act	2003
Planning and Building Act	Law No. 71 of 2008

Pollution Control Act

The purpose of the Pollution Control Act is to protect the external environment from pollution and to reduce existing pollution, as well as to promote better waste management (Article 1).

29 Poland

29.1 General Facts

The national territory covers 312.269 km² and the inhabitants are 38,2 million. The population density is 122/km². In 2009 the GDP per capita amounted to $ 11.287.

Constitutional Structure

The form of government is a parliamentary republic. It is divided into a national level, regional level (voivodship) provincial level (16 provinces), district level (308 districts) and municipal level.

29.2 Production of Raw Materials

Table 107: Production Data – Poland (World Mining Data, Weber and Zsak, 2008)

Resources		2002	2003	2004	2005	2006	Change 02/06	Change 05/06
Nickel	(t)	1300	300	200	300	1600	23,08	433,33
Aluminium	(t)	58.800	57.200	58.900	53.600	55.900	–4,93	4,29
Cadmium	(t)	440	375	356	408	373	–15,23	–8,58
Copper	(t)	502.800	503.200	530.500	511.500	497.200	–1,11	–2,80
Lead	(t)	56.600	54.700	52.700	50.900	50.000	–11,66	–1,77
Zinc	(t)	152.200	153.900	140.300	135.600	126.600	–16,82	–6,64
Gold	(kg)	400	450	600	900	1.300	225,00	44,44
Platinum	(kg)	20	20	20	20	20	0,00	0,00
Palladium	(kg)	12	12	12	12	12	0,00	0,00
Silver	(kg)	1.342.000	1.332.200	1.372.700	1.262.400	1.265.100	–5,73	0,21
Baryte	(t)	30.000	28.000	27.000	25.000	20.000	–33,33	–20,00
Bentonite	(t)	2.800	2.500	2.400	2.200	4.000	42,86	81,82
Feldspar	(t)	85.000	146.000	220.000	300.000	300.000	252,94	0,00

Resources		2002	2003	2004	2005	2006	Change 02/06	Change 05/06
Gypsum	(t)	1.066.000	1.052.000	1.100.000	1.300.000	1.250.000	17,26	–3,85
Kaolin	(t)	252.000	340.000	360.000	380.000	290.000	15,08	–23,68
Magnesite	(t)	24.000	33.000	35.000	37.000	20.000	–16,67	–45,95
Salt	(t)	3.310.000	3.180.000	2.500.000	1.500.000	1.200.000	–63,75	–20,00
Sulphur	(t)	777.000	779.000	820.000	950.000	970.000	24,84	2,11
Steam Coal	(t)	87.839.000	86.727.000	83.486.000	83.632.000	80.650.000	–8,18	–3,57
Coking Coal	(t)	15.876.000	16.147.000	16.526.000	14.071.000	14.570.000	–8,23	3,55
Brown Coal	(t)	58.240.000	60.923.000	61.200.000	61.636.445	60.830.000	4,45	–1,31
Natural Gas	(Mm³)	4.920	4.916	5.000	6.100	6.000	21,95	–1,64
Oil	(t)	437.330	753.260	800.000	800.000	700.000	60,06	–12,50

Table 108: Aggregates Production – Annual Statistics/Poland, 22 April 2008, quantities in million tonnes (UEPG 2008)

Sand & Gravel (1)		Crushed rock (2)		Marine Aggregates		Recycled Aggregates (3)		Manufactured Aggregates (4)	
2005	2006	2005	2006	2005	2006	2005	2006	2005	2006
104,3	115,0	37,7	43,0	n. a.	n. a.	7,2	8,0	1,6	3,0

(1) Sand and Gravel: sold production including crushed grave
(2) Crushed rock: sold production (excluding crushed gravel)
(3) Recycled Aggregates: materials coming from construction and demolition waste used in aggregates market
(4) Manufactured aggregates include blast-furnace-slag, electric-arc-furnace-slag, incinerator bottom ash (IBA), pulverised fuel ash (PFA)

29.3 Normative Basics

29.3.1 Primary Legal Basics

The primary legal basics of mineral extraction activity is the Law of Geological and Mining No. 228 of 2005, latest amendment No 18 of 2009.

29.3.1.1 General Rules

Table 109: Structure of the Geological and Mining Law – Poland

Part 1: General Provisions	
Chapter 1: Scope of application	Article 1–6
Chapter 2: Ownership and Mining Usufruct	Article 7–14
Chapter 3: Concessions	Article 15–30
Part 2: Geological Works	
Chapter 1: Planning and Carrying out of Geological Works	Article 31–39
Chapter 2: Geological Documentation	Article 40–50
Part 3: Exploitation of Minerals	
Chapter 1: Mining Area and Mining Protective Area	Article 51–53
Chapter 2: The Project of Deposit Management	Article 54–56a
Chapter 3: Construction of Mining Plant Facilities	Article 57–62
Chapter 4: Mining Plant Operations	Article 63–79
Chapter 5: Closing Down of the Mining Plant	Article 80–82a
Part 3a: Disposal of Waste in the Subsurface, including Underground Mining Excavations	
	Article 82b – 82f
Part 4: Compensation for Establishment of Mining Usufruct Charges	
	Article 83–87
Part 5: Relations with Neighbours and Liability for Damage	
	Article 88–100
Part 6: Geological Administration Authorities, State Geological Survey and Mining Supervision Authorities	
Chapter 1: Geological Administration Authorities and the State Geological Survey	Article 101–105a
Chapter 2: Mining Supervision Authority	Article 106–117b
Part 7: Penal Provisions	
	Article 118–128
Part 8: Interim and Final Provisions	
Chapter 1: Amendments to Existing Regulations	Article 129–139
Chapter 2: Interim and Final Provisions	Article 140–159
Annex Upper and Lower Limits of Royalty Rates for 2005	

The Law on Geology and Mining distinguishes between basic and common minerals. According to Article 5 basic minerals include:

1) natural gas, oil and its natural derivatives, lignite, hard coal, and coalbed methane,
2) ores of precious metals, metal ores and native metals, including ores of rare and dispersed elements as well as ores of radioactive elements,
3) apatite, baryte, fluorite, phosphate rock, gypsum and anhydrite, pyrite, native sulphur, potassium and potassium/magnesium salts, strontium salts, rock-salt,
4) asbestos, bentonite, diatomite, dolomite, white burnt and stoneware clays, refractory clays and shales, graphite, kaolin, gemstones and decorative stones, quartz, quartzite, magnesite, micas, marbles and crystalline limestones, moulding and glass sands, feldspars, siliceous earth.

All other minerals (i. e. aggregates) are defined as common minerals. A detailed list (including marine raw materials exists in the Law on Mining and Geology as well as in the accompanied regulation 2001/156.

Ownership of mineral rights

The mineral deposits that do not constitute components of land real estate are the property of the State Treasury (Article 7 (1) ML).

Acquiring Mineral Rights

The state has the responsibility about issuing mineral rights. The State Treasury can use mineral deposits as well as dispose of the right thereto by establishing mining usufruct (Article 7 (2) ML). The rights of the State Treasury can be exercised by the authorities that are competent for granting of concessions (3).

Based on the mining usufruct agreement, the mining usufructuary is entitled to, (to the exclusion of other parties), prospect for, explore or exploit a designated mineral (Article 9 ML). According to Article 10 (1) ML the establishment of the mining usufruct takes place by way of an agreement, subject to obtaining a concession. The establishment of the mining usufruct will be preceded by collection of tenders (Article 11 (1) ML). The operator who explored and documented a mineral deposit being the property of the State Treasury and prepared geological documentation with the accuracy required for granting of a concession for mineral exploitation, can demand the establishment of the mining usufruct for its own benefit, with priority over other parties (Article 12 (1) ML).

29.3.1.2 Issuing of Permits

Exploration

A work programme must be enclosed with an application for the granting of a concession for prospecting for and exploration of mineral deposits (Article 19 ML). A concession for prospecting for or exploration of mineral deposits must specify: purpose, scope, type and schedule of the geological works. The surface of the area where the works can be conducted cannot exceed 1200 km² (Article 23 ML).

Exploitation

According to Article 20 ML an application for the granting of a concession for exploitation of minerals must specify: 1) the mineral deposit, that is to be subjected to the exploitation, 2) the scale and manner of the intended mineral exploitation operation, 3) the degree of intended utilization of the mineral deposit, including that of the accompanying minerals, as well as the means enabling the achievement of this aim, 4) the planned location of the mining area and mining protective area and their boundaries.

The following must be appended to the application 1) evidence of the fact the applicant has the right to use the geological documentation with the aim to apply for the concession, 2) a deposit development plan, reviewed by the competent mining supervision authority, 3) evidence of the fact the applicant has the right to the land real estate within the boundaries of which the intended operation of mineral exploitation is to be conducted. In addition to the requirements provided for in Article 22, a concession for mineral exploitation has also designated the boundaries of the mining area and the mining protective area as well as identify the exploitable resources of the mineral deposit and the minimum degree of their utilization (Article 25 ML).

The concession will be granted for a maximum of 50 years. The operator has to provide financial reserves. It is also possible to transfer a concession to another entrepreneur.

The operator who has been granted a concession for the activity referred to in Article 15, is obliged to set up a fund for the closing down of the mining plant (Article 26c 1 ML).

Mining area and mining protective area

The mining area register is maintained by the Minister of Environment (Article 52 ML). The mining area must be delimited for every mineral. The basis for the mining area delimitation is the geological documentation and the deposit development plan (Article 51 ML). The operator applying for a concession for mineral exploitation has to prepare the project of deposit management according to Article 20 (2) ML on the basis of the geological documentation, taking into account the technical and economic factors. The plan must lay down the intentions in the scope of 1) the protection of mineral deposits, including accompanying minerals in the deposit, especially through their comprehensive and effective utilization, 2) the exploitation technology that reduces the adverse impact on the environment (Article 54 ML).

Mining plant operations must proceed on the basis of a mine operating plan (Article 63 ML). According to Article 64 (1) ML an operator must prepare a mine operating plan for each of the mining plants, on the basis of the project of deposit management. The mine operating plan must specify the detailed measures necessary to secure (2):

1) General safety,
2) Fire safety,
3) Work safety and health for employees of the mining plant,
4) Correct and efficient management of the deposit,
5) Protection of the environment and of building facilities,
6) Prevention of damage and its remedy.

A mine operating plan of a plant exploiting a common mineral can be prepared in a simplified form (3). The mine operating plans must be subject to approval by the competent mining supervision authority (4).

Closing down of the mining plant

If a mining plant is closed down, an operator is obliged to (Article 80 ML): 1) secure or close down mining excavations as well as facilities and equipment of the mining plant, 2) secure the unutilized part of the mineral deposit, 3) secure the neighbouring deposits of minerals, 4) take the measures necessary to protect the excavations of the neighbouring mining plants, 5) take measures necessary to protect the environment and reclaim land and develop the post-mining areas.

29.3.1.3 Authorities

The main responsible authority for mining is the Minister of Environment

According to Article 16 ML the Minister of Environment must issue:

1. A concession for an activity in the scope of prospecting for, exploration and exploitation of basic minerals listed in Article 5.

2. Subject to the provisions of paragraphs 1 and 2a, the concessions for prospecting for, exploration or exploitation of basic and common minerals are granted by the voivodeship marshal.
3. Concessions for prospecting for, exploration and exploitation of common minerals, where, at the same time, the following requirements must be met: the mineral exploitation does not exceed 20.000 m³ during a calendar year, the activity does not require the use of explosives to be granted by starost.
4. Granting of concessions for: 1) exploitation of the basic minerals requires consent of the minister responsible for the economy, 2) exploitation of the minerals used for curative purposes require consent of the minister responsible for health, 3) activities carried out within the maritime areas of the Republic of Poland shall require consent of the minister responsible for maritime economy.

29.3.1.4 Fees and Taxes

The amount and manner of payment of the compensation for the mining usufruct must be specified by the agreement referred to in Article 10 ML. The compensation can be paid in a single payment or in instalments. The compensation is the income of the State Treasury (Article 83 ML).

According to Article 84 (1) ML an operator that exploits a mineral from its deposit must pay the royalty for the mineral exploited. The royalty shall be calculated as the product of the rate of the royalty for a given type of mineral and the amount of the mineral exploited in the calculation period (2). The royalty for the accompanying mineral exploited must be calculated as the product of 50% of the amount of the royalty for a given type of mineral and the amount of the accompanying mineral exploited in the calculation period (3).

The royalty must be paid within one month from the end of each quarter of the year. Within the same period, the operator must present to the concession authority and the parties set out in Article 86 ML copies of the receipts of the payments made as well as information containing data specifying the name of the operator, deposit, type, the amount of the mineral exploited in a quarter of the year, the rate adopted and the amount of the royalty determined, the name of the commune, and, in the case when the exploitation is conducted within the area of more than one commune, the amounts of the mineral exploited and the amounts of the royalty due to particular communes.

According to Article 85 ML an operator who has obtained a concession for activities determined in Article 15 paragraph 1 subparagraph 1 and 3 must make the payment arising on prospecting for or exploration of mineral deposits. The concession fee constitutes the product of the fee rate and the number of square kilometres of the area on which the activity is carried out. The amount of the concession fee and the payment dates and manner must be set out in the concession. The operator must submit the copies of the receipts of payments made to the competent authority.

The concession fee rate for the activities consisting in prospecting for mineral deposits for particular groups of minerals per km² shall amount to (state for 2008):
1) energy minerals:
   a) oil, natural gas and coal-bed methane – PLN 103.43,
   b) hard coal – PLN 517.15,
   c) lignite – PLN 206.86,
2) metal ores and native metals – PLN 103.43,
3) chemical minerals – PLN 103.43,
4) rock minerals:
   a) in the land areas – PLN 1.034.29,
   b) in the maritime areas – PLN 103.43.

4. The rate of the concession fee for the activity consisting in mineral deposits exploration and jointly in prospecting and exploration of mineral deposits for particular groups of minerals per km² shall amount to:
1) for energy minerals:
   a) oil, natural gas and coal-bed methane – PLN 206.86,
   b) hard coal – PLN 1.034.29,
   c) lignite – PLN 517.15,
2) metal ores and native metals – PLN 200.86,
3) chemical minerals – PLN 1.034.29,
4) rock minerals:
   a) in the land areas – PLN 10.342.85,
   b) in the maritime areas – PLN 2.585.72.

29.3.2 Additional Legal Basics

Table 110: List of Acts significant for Raw Materials – Poland[1]

Name of Law	No. of Law and Year of Issuing
The Environmental Protection Law	Law No. 25 of 2008
The Act on Providing Information on the Environment and Environmental Protection, Public Participation in Environmental Protection and on Environmental Impact Assessment	Law No. 199 of 2008

---

1 Uberman, R., Ostrega, A (2004), Country Report Poland, in: Department of Mining and Tunnelling, lc.

Name of Law	No. of Law and Year of Issuing
Regulation of the Council of Ministers regarding Payments for the Use of the Environment	No. 55 of 2003
Regulation of the Minister of the Environment, regarding Quality Standards of Soil and Quality Standards of Land	No. 55 of 2002
The Act on the Protection of Agricultural and Forest Land	Law No. 121 of 2004
Regulation of the Minister of Agriculture and Food Economy, regarding Regulation of the Functioning of an Agriculture Land Protection Fund	No.139 of 1998
The Water Law	Law No. 239 of 2005
The Act on Nature Conservation	Law No. 151 of 2009
The Act on Waste	Law No. 39 of 2007
The Act on Preserving the National Character of Strategic Natural Resources of Country	Law No. 97 of 2001
The Act on Land Use Planning and Space Management	Law No. 80 of 2003
The Construction Law	Law No. 156 of 2006
The Act on Liberty of Economic Activity	Law No. 173 of 2004
The Act on Inspection of Environmental Protection	Law No. 112 of 2002
The Ordinance Tax Act	Law No. 8 of 2005
The Act on Local Taxes and Fees	Law No. 9 of 2001
The Act on Real Estate Economy	Law No. 46 of 2000

The Act on Providing Information on the Environment and Environmental Protection, Public Participation in Environmental Protection and on Environmental Impact Assessment

An environmental impact assessment (EIA based on 97/11 EEC) is necessary, if the mining areas are larger than 25 ha as well as the production per year amounts more than 100,000 m^3.

30 Portugal

30.1 General Facts

The national territory covers 92 345 km² and the inhabitants are 10.707.924 (2009 estimation). The population density is 114/km². The GDP per capita for 2009 is estimated at $ 21.407.

Constitutional Structure

The form of government is a parliamentary republic. It is divided into 308 municipalities which are subdivided into more than 4.000 parishes.

30.2 Production of Raw Materials

Table 111: Production Data – Portugal (World Mining Data, Weber and Zsak, 2008)

Resources		2002	2003	2004	2005	2006	Change 02/06	Change 05/06
Iron	(t)	4.500	4.300	4.200	4.100	0	–1,0.00	–1,0.00
Tungsten	(t)	693	715	746	816	780	12,55	–4,41
Copper	(t)	77.227	77.581	95.743	89.541	78.660	1,86	–12,15
Lithium	(t)	191	293	346	311	339	77,49	9,00
Tin	(t)	361	218	220	228	25	–93,07	–89,04
Zinc	(t)	0	0	0	0	7.505		
Silver	(kg)	19.500	21.100	24.400	23.800	20.100	3,08	–15,55
Baryte	(t)	24	25	24	21	24	0,00	14,29
Diatomite	(t)	200	150	120	110	100	–50,00	–9,09
Feldspar	(t)	141.125	126.116	98.262	133.344	125.900	–10,79	–5,58
Gypsum	(t)	834.905	700.000	500.000	500.000	400.000	–52,09	–20,00
Kaolin	(t)	150.193	150.000	180.000	170.000	170.000	13,19	0,00
Salt	(t)	604.969	602.035	661.704	597.945	617.259	2,03	3,23
Talc	(t)	1.328	5.459	6.231	5.362	5.517	3,5.44	2,89
Uranium	(t)	2	0	0	0	0	–1,0.00	

Table 112: Aggregates Production – Annual Statistics/Portugal, 22 April 2008 (UEPG 2008)

Sand & Gravel (1)		Crushed rock (2)		Marine Aggregates		Recycled Aggregates (3)		Manufactured Aggregates (4)	
2004	2006	2004	2006	2005	2006	2005	2006	2005	2006
6,3	10,0 estimated	82,0	87,5 estimated	n. a.	0,0	n. a	n. a.	n. a	n. a.

(1) Sand and Gravel: sold production including crushed gravel
(2) Crushed rock: sold production (excluding crushed gravel)
(3) Recycled Aggregates: materials coming from construction and demolition waste used in aggregates market
(4) Manufactured aggregates include blast-furnace-slag, electric-arc-furnace-slag, incinerator bottom ash (IBA), pulverised fuel ash (PFA)

30.3 Normative Basics

30.3.1 Primary Legal Basics

The primary legal basics of mineral extraction activity is Mining Law No. 18 713 of 1930 as amended by Law No. 90 of 1990.

Furthermore there exist the Ore Deposits Law No. 88 of 1990 and the Quarry Law No. 89 of 1990 as amended by Law No. 270 of 2001. Law 270 of 2001 establishes the legal framework governing the exploration and extracting ("quarrying") of mineral masses (i. e. land owner minerals).

30.3.1.1 General Rules

Table 113: Structure of the Mining Law of 1990.

Chapter I Basic Arrangements		
Section I Ownership of the Deposits and Mineral Excavations and their Classification	Article 1	1 §
	Article 2	
	Article 3	Item 1–3
	Article 4	1 §
Section II Defined Areas	Article 5	§§ 1–2

Section III Services and the concerned Research Teams as Assistance for the Mining Work	Article 6	1 §
Chapter II Notification and registration of the discovery of mineral excavations		
Section I Announcement	Article 7	
	Article 8	Item 1–4 1 §
Section II Registering of the Announcement	Article 9	
	Article 10	Item 1–2 §§ 1–4
	Article 11	Item 1–4
	Article 12	
	Article 13	§§ 1–2
	Article 14	
	Article 15	
Section III Transmission of the Rights of a Mining Announcement Checked by a Register	Article 16	
Section IV Invalidity and Obsolescence of the Mining Registers	Article 17	Item 1–8
Chapter III Inquiries		
Section I Procedure of Inquiries	Article 18	
	Article 19	Item 1–4
	Article 20	1 §
Section II Conditions in which the Procedure of Inquiry can be Effectuated	Article 21	1 §
	Article 22	Item 1–5
	Article 23	
	Article 24	1 §
	Article 25	
	Article 26	Item 1–4
Section III Exemption of Taxes and Contributions	Article 27	

Chapter IV Mining Concession

Sector I Request for Concession	Article 28	§§ 1–2
	Article 29	Item 1–3, 1 §
	Article 30	Item 1–11, 1 §
	Article 31	1 §
	Article 32	1 §
Section II Recognition and Demarcation	Article 33	§§ 1–3
	Article 34	1 §
	Article 35	Item 1–11
	Article 36	1 §
	Article 37	
Section III Transfer of the Rights to the Concession	Article 38	Item 1–2 §§ 1–2
Section IV Permit of Concession	Article 39	1 §
	Article 40	§§ 1–2
	Article 41	
	Article 42	
	Article 43	
Section V Concession of the Mining Protectorates	Article 44	§§ 1–2
	Article 45	
	Article 46	
	Article 47	
Section IV Fragmentation and Reduction of the Concessions Area	Article 48	
	Article 49	§§ 1–2
Section VII Recognized Contracts for Mining Concessions	Article 50	
	Article 51	Item 1–3
	Article 52	
	Article 53	
	Article 54	Item a–b

Chapter V Exploration of the mining concessions

Section I General Arrangements	Article 55	§§ 1–2
	Article 56	§§ 1–3

Section II Obligations and Rights of the Mining Concessionaires	Article 57	Item 1–17 1 §
	Article 58	§§ 1–3
	Article 59	1 §
	Article 60	Item 1–2
Section III Technical Management of the Mining Works	Article 61	§§ 1–3
	Article 62	
	Article 63	1 §
Section IV Supervising of the Mining Concessions	Article 64	
Section V Mining Statistic	Article 65	
	Article 66	
Chapter VI Police force and relative jurisdiction for the mining concessions		
Section I General Arrangements	Article 67	
	Article 68	
	Article 69	§§ 1–3
	Article 70	
	Article 71	
Section II Procedures with Health Risk	Article 72	1 §
Section III Procedures for Damages caused by Third Parties	Article 73	1 §
	Article 74	§§ 1–2
Section IV Working Accidents	Article 75	
Chapter VII The accidents of work will be regulated by the respective legislation		
	Article 76	
	Article 77	§§ 1–6
	Article 78	
	Article 79	1 §
	Article 80	
	Article 81	§§ 1–3
	Article 82	§§ 1–6
Chapter VIII Applicable penalties to the mining concessionaires		
	Article 83	Item a–b

	Article 84	1 §
	Article 85	Item 1–4
	Article 86	1 §
	Article 87	
	Article 88	§§ 1–2
	Article 89	
Chapter IX Abandonment of the mining concessions		
	Article 90	Item 1–2
	Article 91	
	Article 92	§§ 1–3
	Article 93	§§ 1–2
	Article 94	1 §
	Article 95	Item 1–4
	Article 96	1 §
	Article 97	§§ 1–2
	Article 98	Item 1–2
	Article 99	1 §
	Article 100	
Chapter X Taxes		
	Article 101	§§ 1–3
	Article 102	
	Article 103	
	Article 104	Item 1–2 §§ 1–2
	Article 105	
	Article 106	
	Article 107	§§ 1–2
	Article 108, Article 109	1 §
	Article 110	§§ 1–2
	Article 111	
	Article 112	§§ 1–2

Chapter XI General and transitory arrangements		
	Article 113	
	Article 114	
	Article 115	1 §
	Article 116	
	Article 117	1 §
	Article 118	
	Article 119	
	Article 120	
	Article 121	§§ 1–2
	Article 122	
	Article 123	
	Article 124	
	Article 125	
	Article 126	
	Article 127	1 §
	Article 128	
	Article 129	
Annex	**Example A – Example I**	
Table 1	Taxes	
Table 2	Fines	

Minerals classification

The minerals are divided under the following three classes (Art. 3 ML):
1. metalliferous deposits,
2. non-metalliferous deposits as graphite, coal, lignite, peat, talc, rock salt, potassium, phosphates, kaolin,
3. hydrocarbons and bituminous substances.

Ownership of mineral rights

Furthermore, Law No. 90 of 1990 classifies mineral occurrences into two categories:[1]

Minerals "deposits": These encompass all minerals considered to be of importance to the Portuguese economy, by virtue of their rarity, high specific

---

1 Department of the Environment (1994), lc.

importance for industry. The ownership of mineral rights to mineral deposits belongs to the State.

Mineral "masses": These include natural stones exploited in quarries, sand, clay and other materials not classified as mineral deposits. The ownership of mineral rights to mineral masses belongs to the landowner.

The ownership of mineral rights of minerals deposits belong to the state. The exploitation of these deposits may be granted for mining work (Article 1 ML). The quarries ('mineral masses') of any kind can be exploited by the landowner (Article 2 ML).

Acquisition of mineral rights

The Minister of Economics (Directorate General of Mines and Geological Services) grants the mineral right over State-owned mineral resources by contract, one for prospecting and exploration, and another for exploitation (mining). In both cases the contract follows an application submitted by any natural or legal person, or it is based on a public tender. Each contract for prospecting and exploration or for exploitation covers a particular geological resource in a well-delimited area, as specified in the contract. The applicant must give details about its professional skills or experience to conduct the operations he applied for, and about his financial capabilities to mobilize the necessary funds.[2]

30.3.1.2 Issuing of Permits

Exploration

Once the rights for prospecting and exploration are granted, the competent authority (Directorate General of Mines and Geological Services) notifies the applicant for the conclusion of the respective contract. This contract includes provisions about:[3]

- delimitation of the exploration area
- type of mineral deposits covered by the contract
- contract duration and extensions
- conditions of gradual abandonment of the exploration area
- work program and expenditures
- regularity presentation of working programs and progress reports

---

2 Magna, C. (2004), Country Report Portugal, in: Department of Mining and Tunnelling, lc.
3 Ibidem.

- value of the definitive bond to secure the fulfilment of the contractual obligations

Within the scope and the duration of the contract for prospecting and exploration, the holder is entitled to carry out the all necessary studies and works.

Exploitation

To get an exploitation concession within an exploration contract area, the exploration holder must submit an application to the competent authority (Directorate General of Mines).

According to Article 29 (ML) the following is required from the applicant:

1. The nationality, profession and residence of the owner of the manifest;
2. The place/site (county and district) where the deposit has been discovered;
3. Description of the deposit;

Additionally the following documents are required (Article 30 ML):

- A certificate of registration referred to (3) Article 11 ML;
- The receipt of the required payment to the public treasury;
- A plan describing existing topographical conditions (scale 1 : 10.000);
- A work plan (scale 1 : 1.000);
- Measures related to environment protection, land recovery and pre-viability study of the exploitation:
  o A description of mining operation and estimated budget including deposit conditions according carried out exploration;
  o The operation method that will be taken;
- The applicant must prove the necessary capital needed for the proposed work.

The Ministry, following a recommendation made by the Directorate General of Mines and Geological Services, can also order a public invitation for prospecting and exploration activities proposals for defined resources, through a public tender. A tender can also be promoted for the direct granting of an exploitation contract. In that way, the invitation must be published in the Diário da República.

Consultation process

After submission the public inquiry starts for 30 days including a consultation process with the relevant land use planning authority and environment authority. When the public inquiry is finished, the Directorate General of Mines and Geological Services must submit the application to the Minister of

Economics which will issue (or refuse) the concession. A concession contract includes: granted area delimitation, mineral deposit indication whose exploitation is granted, concession duration and conditions required for extensions, indication of rights and obligations

After completion of the grant, the Directorate General of Mines and Geological Services will send it to its Board of Governors. The Board shall provide its opinion in order to enable the Minister of Commerce and Communications to accept or reject the concession (Article 39 ML). If it is to initiate a new exploration of deposits, this can only be authorized by the Minister of Trade and Communications. As soon as the Minister of Trade and Communications has made its decision on the granting process, the Directorate General of Mines and Geological Services must notify the respective person not later than 90 days from the date of the order. (Article 40 ML).

Mining concessions may not be awarded without the authorization of the Minister of Trade and Communications (Article 50). According to Article 57 ML the concessionaire must start mining works within 3 months from the date of the concession receipt; carry out the works based on an (approved) mining plan; establish the works necessary for safety and health.

Operators are required to submit each year to the Directorate General of Mines and Geological Services a note of the work performed the previous year, statistical data and any other documents that are required (Article 65 ML).

Quarry Law 270 of 2001

The exploration or the exploitation of mineral masses can only be realized under an exploration permit or exploitation permit under the terms and conditions established in the Quarry Law. The exploration permit has the initial period of 6 months, counted from the date of its attribution. It can be extended for another period of 6 months, if the permit holder so requests within the 30 days before the ending of the initial period. The applicant of an exploitation permit must present to the competent authority, the following documents:

- A Property Land Certificate or contract certificate when the explorer will not be the landowner
- A quarry work program (including a restoration plan) and details about the technical manager responsible by the operations
- An Impact Environmental Study, only if the quarry operation is subjected to an EIA
- A localization plan at 1:25.000 scale indicating the site accesses within a radius of 4 km

- A map at 1:2.000 scale with quarry borders, access roads, and the neighbouring landowners.
- A topographical plan at 1:500 or 1:1.000 scales indicating the localization of the quarry industrial facilities.
- An economic pre-feasibility study

30.3.1.3 Authorities

The main responsible authority for mining is the Minister of Economics (Directorate General of Mines and Geological Services).

All mining concessions will be inspected at least once each year. The technical officer must specify clearly any defects noticed, with express indication how to correct it (Article 64 ML). The competent authority of the respective mining district must report each year the Director General of Mines and Geological Services the state of work. Article 92 ML requires (in all cases mentioned in Article 90) the General Department of Mines and Geological Services to inspect the mines and to indicate security measures. Where it is necessary to adopt security measures, the Directorate General of Mines and Geological Services must report back to the competent authority (local level); to this order the operator must to proceed with its implementation within the period determined. Completion of work or lack of initiation will be reported to the Directorate General of Mines.

30.3.1.4 Fees and Taxes

According to Article 101 ML the operator is required to pay to the State a fixed tax dependent on the concession area and a charge proportional to the amount of useful mineral substance extracted during each calendar year.

Table 114: Taxes in accordance with Article 66 of Decree No. 88 of 1990 (Magno, 2004)

Articles	Assignments	Taxes
Article 8	Prospecting and exploration contract	
	Metallic minerals	€ 150
	Non metallic minerals	€ 75
File d) n 1 of Article 8	Exploration period	€ 50

Articles	Assignments	Taxes
Article 11	Transmission exploration contractual position.	
	Metallic minerals	€ 125
	Non metallic minerals	€ 50
Article 20	Experimental concession contract	€ 200
Article 21	Concesión contract.	€ 325
File c) Article 21	Registering new substance to the concession contract.	€ 150
File d) Article 21	Concession contract extension	€ 50
Article 22	Concession transmission	€ 250
Article 23	Concession landmark area	€ 150
Article 24	Alteration of concession area	€ 150
Article 25	Voluntary integration of concessions for each concession.	€ 150
N 3 Article 27	Work Plan alteration	€ 100

For the exercise of mineral masses exploitation activity according to the use Article 56. Decree No. 89 of 1990:

Table 115: Articles, Assignments and Taxes according to Article 56 Law No. 89 of 1990 (Magno, 2004)

Articles	Designation	Taxes
File a) n 2 Article 18	Establishment permit attribution	€ 100
File b) n 2 Article 18	Establishment permit attribution	€ 250
File c) n 2 Article 18	Establishment permit attribution	€ 375
File a) n 2 Article 20	Work plan alteration	€ 75
Article 26	Establishment permit transmission	

30.3.2 Additional Legal Basics

Table 116: List of Acts significant for raw materials – Portugal (Magno, 2004)

Name of Law	No. and Year of Issuing
Environmental Impact	Decree Law No. 186 of 1988
Noise Standards	Decree Law No. 292 of 1989
Water Quality Standards	Decree Law No. 74 of 1990
Air Quality Standards	Decree Law No. 352 of 1990
Industrial Licensing	Decree Law No. 109 of 1991, 282 of 1993
Civil Code	
Commercial Company Law	Decree Law No. 76-A of 2006 last amended Decree Law No. 357-A of 2007
Legal framework of environment	Law No. 11 of 1987
Legal framework of Geological Resources	Decree Law No. 90 of 1990
Mineral Deposits Law	Decree Law No. 88 of 1990
Quarrying Law	Decree Law No. 270 of 2001
Legal framework of spatial planning	Law No. 48 of 1998
Management of the territory and planning system	Decree Law No. 380 of 1999
Wastes from extractive industries	Decree Law No. 544 of 1999
Environment Impact Assessment	Decree Law No. 69 of 2000

31 Romania

31.1 General Facts

The national territory covers 238,391 km² and the inhabitants are 21,6 million. The population density is 91/km². In 2009 the GDP per capita amounted to $ 7.542.

Constitutional Structure

The form of government is a unitary semi-presidential republic. Romania is divided into 41 districts.

31.2 Production of Raw Materials

Table 117: Production Data – Romania (World Mining Data, Weber and Zsak, 2008)

Resources		2002	2003	2004	2005	2006	Change 02/06	Change 05/06
Manganese	(t)	12.000	15.305	16.617	18.000	13.673	13,94	−24,04
Aluminium	(t)	187.100	196.800	222.300	243.600	258.300	38,05	6,03
Copper	(t)	19.300	23.400	20.400	16.300	12.200	−36,79	−25,15
Lead	(t)	15.100	15.700	18.300	11.610	7.500	−50,33	−35,40
Zinc	(t)	23.500	21.260	23.600	13.784	9.600	−59,15	−30,35
Gold	(kg)	1.137	1.878	1.430	500	500	−56,02	0,00
Silver	(kg)	18.000	18.000	18.000	18.000	18.000	0,00	0,00
Baryte	(t)	1.000	208	73	60	21	−97,90	−65,00
Bentonite	(t)	15.402	17.637	18.161	18.190	14.600	−5,21	−19,74
Diatomite	(t)	20.128	31.298	14.192	13.000	11.255	−44,08	−13,42
Feldspatr	(t)	50.864	71.717	60.636	74.927	55.300	8,72	−26,19
Graphite	(t)	1.001	1.014	395	486	350	−65,03	−27,98
Kaolin	(t)	22.514	21.724	22.337	26.772	22.500	−0,06	−15,96
Salt	(t)	2.200.000	2.415.274	2.398.607	2.550.000	2.612.749	18,76	2,46
Sulphur	(t)	20.000	18.464	17.836	16.000	8.437	−57,82	−47,27

Resources		2002	2003	2004	2005	2006	Change 02/06	Change 05/06
Talc	(t)	7.292	10.082	9.725	6.760	10.745	47,35	58,95
Steam Coal	(t)	3.000.000	3.020.909	2.675.737	3.000.000	2.841.024	−5,30	−5,30
Brown Coal	(t)	30.189.000	30.029.987	30.410.351	31.122.000	32.600.000	7,99	4,75
Natural Gas	(Mm³)	13.425	13.026	13.000	12.500	12.000	−10,61	−4,00
Oil	(t)	5.840.000	5.661.500	5.500.000	5.600.000	5.500.000	−5,82	−1,79
Uranium	(t)	106	106	106	106	106	0,00	0,00

Table 118: Aggregates Production – Annual Statistics/Romania, 22 April 2008 (UEPG 2008)

Sand & Gravel (1)		Crushed rock (2)		Marine Aggregates		Recycled Aggregates (3)		Manufactured Aggregates (4)	
2005	2006	2005	2006	2005	2006	2005	2006	2005	2006
n. a.	15,5	n. a.	6,5	n. a.	0,0	n. a	0,5	n. a.	0,5

(1) Sand and Gravel: sold production including crushed gravel
(2) Crushed rock: sold production (excluding crushed gravel)
(3) Recycled Aggregates: materials coming from construction and demolition waste used in aggregates market
(4) Manufactured aggregates include blast-furnace-slag, electric-arc-furnace-slag, incinerator bottom ash (IBA), pulverised fuel ash (PFA).

31.3 Normative Basics

31.3.1 Primary Legal Basics

The primary legal basics of mineral extraction activity is the Mining Law No. 85 of 2003.

31.3.1.1 General Rules

Table 119: Structure of the Mining Law – Romania

Chapter I:	General Provisions Articles 1–5
Chapter II:	Right to use and the Access to the Land on which Mining Activities are performed Articles 6–12
Chapter III	Turning Mineral Resources into Value Articles 13–37
Chapter IV:	Rights and Obligations of the Title Holder Articles 38–39
Chapter V:	Authorizations Articles 40–43

Chapter VI:	Mining Fees, Taxes and Royalties Articles 44–50
Chapter VII:	Closure of Mines Articles 51–53
Chapter VIII:	The Competent Authority Articles 54–55
Chapter IX:	Responsibility of the Line Ministry Article 56
Chapter X:	Sanctions Articles 57–59
Chapter XI:	Transitory and Final Provisions Articles 60–69

The mineral resources which are the subject of the Mining Law are (Article 2 ML) include coal, ferrous and non ferrous ores, aluminum rocks and minerals, noble, radioactive, rare and disperse metals, salts, non metallic useful substances, useful rocks, precious and semiprecious stones, peat, bituminous rocks.

Ownership of mineral rights

The mineral resources located on the territory and in the subsoil of the country and of the continental shelf in the Romanian economic area of the Black Sea, delimited in accordance with the principles of international law and of international regulations to which Romania is a party, are the exclusive object of public property and they belong to the Romanian State (Article 1 ML). The State has the responsibility about issuing mineral rights. Mining activities can be performed by natural persons or legal entities (registered according Rumanian law).

31.3.1.2 Issuing of Permits

Prospection (Article 14 MG)

Prospecting can be conducted on the basis of a non-exclusive permit, issued by the competent authority, within a perimeter, defined by topographic coordinates (Article 14 (1) ML). The shape and size of the prospecting perimeter are established by the competent authority. The prospecting permit is issued for a period of maximum 3 years, without a renewal right, with an annual advance payment of a fee for prospecting activity (Article 14 (2) ML).

Title holders of prospecting permits are obliged to carry out an annual program, having a minimum value negotiated with the Competent authority, correlated with the validity period of the permit and the size of the prospecting perimeter, at the time the prospecting permit is issued (Article 14 (3) ML).

The title holder of the prospecting permit must present to the Competent Authority annual reports regarding the work conducted. The title holder presents a final report, comprising the investigation method used, the work conducted, justification of the related expenditures and the results obtained in no more than 60 days from the expiry of the period for which the permit has been issued (Article 14 (4) ML).

According to Article 14 (5) ML the title holder of a prospecting permit that takes part in a public offering, organized for the award of an exploration license, within the perimeter it carried out prospecting works, benefits from the bonus set out in Article 15 (8).

Exploration (Articles 15–17 ML)

According to Article 15 exploration can be conducted on the basis of an exclusive permit (1). The exploration permit will be granted to the winner of a public offering, organized by the Competent Authority. The initiative for the concession of exploration mining activities may belong to the Competent Authority or the interested Romanian/foreign legal persons (3). The list of exploration perimeters is established by the Competent Authority, through an order, which is published in the Official Monitor of Romania, part I. (4). In order to participate in the public offering, the Romanian/foreign legal persons must submit offers within the period established by the Competent Authority (5). The offers must contain the proposed exploration program, the documents regarding the technical and financial capabilities of the applicant (6). The proposed exploration program includes the annual exploration amount of works and the related expenditures, which are mandatory (7). The conditions regarding the organization of the public offering, the bonus granted to a prospecting permit Title holder, must be established by the Competent Authority(8).

The exploration permit will be granted for a maximum period of 5 years, with a renewal right of no more than 3 years (Article 16 (1) ML). The title holder has the right to reduce the area related to the exploration permit, with the Competent Authority agreement, on the basis of phase reports, proving that all the necessary environmental works were executed, having the obligation to carry out the exploration works for the first year (5).

Exploitation (Articles 18–37 ML)

Mining can be conducted on the basis of an exclusive permit, granted in accordance with this law (Article 18 ML). The issuing of an exploitation license can be done in different ways: It will be granted: a) Directly to the title holder of the exploration permit, on its request for any of the mineral resources discovered, in maximum 90 days from the exploration final report submission, satisfactory to the Competent authority; b) the winner of a public offering, organized by the Competent Authority.

The list of exploitation perimeters is established by order of the Competent Authority, which will be published in the Official Gazette of Romania (Article 19 – (2) ML). In order to participate in the public offering, the Romanian/foreign legal persons must submit offers within the period established by the Competent Authority (Article 19 (3) ML). The offers must contain the documents regarding the technical and financial capabilities of the applicant, as well as any other documents established by the Competent Authority through the bidding procedure for the public offering (Article 19 (4) ML).

According to Article 20 – (1) the exploitation permit will be granted based on an application, accompanied by:

(a) a feasibility study regarding the capitalization of the mineral resources and the deposit protection;
(b) a development plan of the exploitation;
(c) the environmental impact study;
(d) an environmental rehabilitation plan;
(e) a social impact assessment and social mitigation plan.

The exploitation permit shall be granted for maximum 20 years, with the right of continuation for successive periods of 5 years each (Article 20 (2) ML). The titleholder of the exploitation permit must pay annually a tax on exploitation activity and a mining royalty (3). The title holder has to establish a financial guarantee for environmental rehabilitation(4).

The title holder of a permit can transfer its rights to another legal person only subject to the prior written agreement issued by the Competent Authority (Article 24 (1) ML). Articles 28–29 ML regulate the extraction activities related to construction minerals. The permit will be issued to natural or legal persons for a period of one year. The exploitation permit will be released to the first applicant (Article 28 (1) ML).

Rights and obligations of the title holder

The title holder of the permit has the following rights (Article 38 ML):

(a) to access to land needed for the conduct of mining activities within the boundaries of the perimeter provided for by the license/permit;
(b) to conduct of all mining activities provided by the license/permit;
(c) to dispose of the quantities of mineral resources produced;
(d) to manage sources of surface and underground water as necessary to conduct the mining activities;
(e) to obtain from the Competent Authority data and information required for the conduct of mining activities, to keep and use such data and infor-

mation, as well as those obtained from its own operations, for the entire duration of the permit;

(f) to build roads, bridges, railways, electricity networks and other infrastructure utilities necessary for the conduct of mining activities, according to the law.

The title holder of the permit has the following obligations (Article 39 (1) ML):

(a) to prepare before commencement and during execution of mining activities technical and economic documentation for carrying out the mining activities, documentation for environmental protection cleared in accordance with the Environmental Protection Law, according to Article 22d ML;
(b) to start mining activities within no more than 210 days from the effective date of the permit;
(c) to obtain, prepare, keep up to date and to submit to the Competent Authority, on the scheduled dates, all data, information, and documentation established in the permit, concerning the mining activities carried out and the results obtained in order to be register in the Mining Book and the Mining Cadastre;
(d) to inform the competent authority about the control inspections made by the local environmental and labour protection authorities;
(e) to regularly update the mining activity cessation plan and submit it to the Competent Authority for approval;
(f) to execute and finalize the environmental rehabilitation of the perimeters affected by the mining works performed.

31.3.1.3 Authorities

The main responsible authority for mining is the Minister of Environment (National Agency for Mineral Resources).

The National Agency for Mineral Resources represents the interests of the state in the sector of mineral resources, according to the competences provided by this law (Article 3 ML).

31.3.1.4 Fees and Taxes

According to Article 44 (1) ML the title holder of the permits must pay to the State budget a tax for the activity of prospecting, exploration and exploitation of mineral resources and a mining royalty. It is doubled after two years and becomes five times greater after four years. The taxes set out in para. (2) – (4)

are due annually and must be paid in advance in respect of the next year until the 31 December of the current year (6).

The mining royalty, owed to the State budget, must be equivalent to a percentage quota of the value of the mining production, as follows (Article 45 (1) ML):

a) 2 % for coal, ferrous and non ferrous ores, aluminum rocks and minerals, noble, radioactive, rare and disperse metals, precious stones and gems;
b) 6 % for non metallic useful substances;
c) 6 % for useful rocks, except for ornamental rocks for which the quota is of 10 %;
d) 8 % for salts.

The value of the mining production does not include the costs of processing of the extracted products (3). The mining royalty is due beginning with the day of commencement of production and is payable quarterly until the 20th day of the first month of the following quarter (4).

31.3.2 Additional Legal Basics

Table 120: List of Acts significant for Raw Materials – Romania

Name of Law	No. of Law and Year of Issuing
Forest Code	Law No. 26 of 1996
Civil Code	
Law on Fiscal Code of Romania	Law No. 571 of 2003
Norms for applying the Mining Law	No. 85 of 2003
Law on Drinking Water	Law No. 458 of 2002
Law on the Environmental Protection	Law No. 137 of 1995
Water Law	Law No. 107 of 1996, last amendment by Law No. 310 of 2004
Law on Land Resources	1991
Law regarding the Protection of the National Heritage	Law No. 182 of 2000
Law on Public Property and its juridical regime	Law No. 213 / 1998
Law on Local Public Administration	Law No. 215 / 2001
Law on Trading Companies	Law No. 31 / 1990

Name of Law	No. of Law and Year of Issuing
Law on the Protection of Historical Monuments	Law No. 422 / 2001
Law on Regional Development in Romania	Law No. 315 / 2004
Decision on the minimum requirements for improving the safety and health protection of workers in surface and underground mineral-extracting industries	
Decision concerning the minimum requirements to safeguard the safety and health of workers in the mineral-extracting industries through drilling	

Law on Environmental Protection

The object of the law is to regulate environmental protection, an objective of major public interest, on the basis of the principles and strategic elements which lead to the sustainable development of society (Article 1). The central environmental protection authority, in consultation with the competent ministries, must establish the permitting procedure regarding the environmental protection issues related to mining activities (Article 48).

Water Law

The provisions of the Water Law have the following objectives: a) the conservation, development and protection of water resources, as well as the ensuring of a free water flow; b) the protection against any form of pollution and modification of the characteristics of the water resources, of their banks and beds, or basins (Article 2). For the designing of surface mining activities that can influence the ground water reserve or water supply, approbriate rehabilitation and flood protection measures must be proposed (Article 10). The extraction of the mineral aggregates is allowed only from the evaluated reserves (granted from the Ministry of Waters, Forests and Environmental Protection), under the conditions for water flows and river beds and banks stability, and by taking care not to affect the structures in the areas directly or indirectly influenced by the water flow regime (Article 33).

32 Russian Federation

32.1 General Facts

The Russian Federation has a national territory of 17.075.400 million km² and the inhabitants are 142 million people (8.3/km²). There are 128 nationalities living in Russia. The GDP per capita in 2009 is amounted to $ 8.693.

Constitutional Structure

The Constitution of 1993 declared the Russian Federation into a presidential republic with a parliament (Federalnoe Sobranie), also called Federal Assembly. Russia is divided into 83 federal areas, which are grouped into seven federal districts.

32.2 Production of Raw Materials

Table 121: Production Data – Russian Federation (European Section) (World Mining Data, Weber and Zsak, 2008)

Resources		2002	2003	2004	2005	2006	Change 02/06	Change 05/06
Gold	(kg)	11.788	11.905	11.851	11.423	11.154	–5,38	–2,35
Platinum	(kg)	5.400	5.600	5.600	5.800	5.800	7,41	0,00
Silver	(kg)	40.000	70.000	106.200	110.000	110.000	175,00	0,00
Asbestos	(t)	600.000	616.000	720.000	700.000	680.000	13,33	–2,86
Baryte	(t)	5.000	4.800	6.000	6.000	6.300	26,00	5,00
Bentonite	(t)	800.000	720.000	696.000	680.000	664.000	–17,00	–2,35
Feldspar	(t)	34.000	33.200	32.000	31.200	28.000	–17,65	–10,26
Fluorspar	(t)	32.000	20.000	17.000	17.000	21.000	–34,38	23,53
Graphite	(t)	5.900	5.700	5.500	5.300	5.100	–13,56	–3,77
Gypsum	(t)	520.000	550.000	600.000	800.000	1.200.000	130,77	50,00
Kaolin	(t)	1.500.000	1.300.000	1.200.000	1.150.000	1.000.000	–33,33	–13,04
Magnesite	(t)	1.350.000	1.080.000	900.000	885.000	1.080.000	–20,00	22,03
Perlite	(t)	60.000	58.000	56.000	55.000	52.000	–13,33	–5,45
Phosphate	(t)	1.614.450	1.660.890	1.701.795	1.692.960	1.590.000	–1,51	–6,08
Potash	(t)	4.400.000	4.650.000	4.900.000	5.000.000	4.000.000	–9,09	–20,00

Resources		2002	2003	2004	2005	2006	Change 02/06	Change 05/06
Salt	(t)	1.800.000	2.500.000	3.000.000	2.800.000	2.700.000	50,00	−3,57
Sulphur	(t)	4.100.000	4.200.000	4.400.000	7.100.000	7.000.000	70,73	−1,41
Talc	(t)	95.000	85.000	80.000	45.000	50.000	−47,37	11,11
Vermiculite	(t)	25.000	23.000	21.000	25.000	25.000	0,00	0,00
Brown Coal	(t)	7.600.000	7.000.000	7.030.000	7.090.000	9.000.000	18,42	26,94
Natural Gas	(Mm³)	535.500	531.000	540.000	549.000	558.000	4,20	1,64
Oil	(t)	329.400.000	360.000.000	378.000.000	405.000.000	431.872.938	31,11	6,64
Oil Shales	(t)	1.500.000	1.400.000	1.300.000	1.700.000	1.900.000	26,67	11,76

Table 122: Production of lime – Russian Federation. (Metric tons), (USGS, 2008)

	2002	2003	2004	2005	2006
Lime (industrial and construction)	8.000 000	8.000 000	8.200 000	8.200 000	8.200 000

32.2.1 Primary Legal Basics

The primary legal basic of mineral extraction activity is the Mining Law ("Law on Subsoil") No. 1992 of 2395 as amended in 2007.

32.2.1.1 General Rules

Table 123: Structure of the Law of the Russian Federation on Subsoil

Section I General Provisions	Article 1. Russian Federation Legislation on Subsoil
	Article 1.1. Statutory Regulation of Relations Pertaining to the Use of Subsoil
	Article 1.2. Title to Subsoil
	Article 2. State Subsoil Reserve
	Article 2.1. Areas of Subsoil under Federal Jurisdiction
	Article 2.2. Federal Fund of Reserve Areas of Subsoil
	Article 3. Powers of Federal Government Bodies in Governing Relations with Respect to the Use of Subsoil
	Article 4. Powers of the Government Bodies of the Subjects of the Russian Federation in Governing Relations Pertaining to the Use of Subsoil
	Article 5. Powers of Local Self-Government Bodies in Governing Relations Pertaining to the Use of Subsoil

Section II **Use of Subsoil**	Article 6. Types of Use of Subsoil
	Article 7. Subsoil Areas Granted for Use
	Article 8. Restrictions on the Use of Subsoil
	Article 9. Subsoil Users
	Article 10. Terms of Use of Subsoil Plots
	Article 10.1. Grounds for Awarding the Right to Use Subsoil Plots
	Article 11. Grounds for Awarding the Right to Use Subsoil Plots
	Article 12. Contents of the Permit for the Use of Subsoil
	Article 13. Abolished
	Article 13.1. Tenders and Auctions for the Right to Use Subsoil Plots
	Article 14. Rejection of Tender or Auction Bids or an Application to Obtain the Right to Use Subsoil on a Non-Competitive Basis
	Article 15. Terms of Use of Subsoil Plots
	Article 16. Organisational Support to the State Licensing System
	Article 17. Anti-monopoly Provisions for the Use of Subsoil
	Article 17.1. Assignment of Rights for the Use of Subsoil and Renewal of the Effective Permit
	Article 18. Granting of Subsoil Areas for Development of Fields Containing Commonly Occurring Minerals
	Article 19. Production of Commonly Occurring Minerals by Owners of Land plots, Land Users, Land Possessors and Tenants
	Article 19.1. Exploration and Production of commonly occurring Minerals and Groundwater by Users of Subsoil
	Article 20. Grounds for Termination of the Right for the Use of Subsoil
	Article 21. Procedures for Early Termination of the Right for the Use of Subsoil
	Article 21.1. The Use of Subsoil Plots in the Case of Early Termination of the Right to Use Subsoil Plots
	Article 22. Basic Rights and Obligations of the Users of Subsoil

Section III Rational Use and Conservation of Subsoil	Article 23. Basic Provisions for Rational Use and Conservation of Subsoil
	Article 23.1. Geological-and-Economic and Cost Evaluation of Mineral Deposits and Subsoil Plots
	Article 23.2 Procedures for Mineral Field Development and the Use of Subsoil for Purposes Unrelated to Mineral Production
	Article 23.3 Primary Processing of Mineral Raw Materials by the Users of Subsoil
	Article 24. Basic Provisions for Safety of Operations Related to the Use of Subsoil
	Article 25. Terms and Conditions for Construction Within Areas Containing Mineral Reserves
	Article 25.1. Granting and Taking Land Plots when Works Are Performed in Connection with a Geological Survey and Another Use of the Sub-Soil
	Article 26. Abandonment and Conservation of Mineral Production Enterprises and Underground Facilities Unrelated to Mineral Production
	Article 27. Geological Information on Subsoil
	Article 28. State Records and Registration
	Article 29. State Evaluation of Mineral Reserves
	Article 30. State Cadastre of Fields and Occurrence of Mineral Reserves
	Article 31. State Balance of Mineral Reserves
	Article 32. Maintenance of the State Cadastre of Fields and Occurrences of Mineral Reserves and the State Balance of Mineral Reserves
	Article 33. Conservation of Subsoil Areas of Special Scientific or Cultural Value
	Article 34 Rewards for Mineral Field Discovery
Section IV State Regulation of the Use of Subsoil	Article 35. Objectives of State Regulation of the Use of Subsoil
	Article 36. State Control over Relations in the Use of Subsoil
	Article 36.1. State Geological Exploration
	Article 37. State Control over the Rational Use and Conservation of Subsoil
	Article 38. State Inspection of Operating Safety with Respect to the Use of Subsoil

Section V Payments Required in Subsoil Use	Article 39. Subsoil Use Payment System
	Article 40. One-Off Subsoil Use Payments in the Case of Onset of Certain Events Stipulated by a Permit
	Article 41. Payment for the Provision of Geological Information on Subsoil
	Article 42. The Fee Charged for Participation in a Tender (Auction)
	Article 43. Regular Payments for the Use of Subsoil
Section VI Responsibility for Breaking This Law	Article 49. Responsibility for Breaking this Law
	Article 50. Dispute Resolution
	Article 51. Damage Compensation
Section VII International Agreements	Article 52. International Agreements

Ownership of Mineral rights

The State has the responsibility to issue mineral rights. These can principally be transferred to natural or legal persons. At the same time, the property rights to strategic raw materials can be limited.[1] Uranium, diamonds, high-quality quartz, the element group yttrium, nickel, cobalt, tantalum, niobium, beryllium, lithium and the platinum group are counted to the strategic raw materials.

32.2.1.2 Issuing of Permits

Article 6 ML provides basic regulations for the exploration and extraction of mineral resources and construction of underground facilities. The issuing of the permit rights can be done for the exploration and production separately or together.

The permit holder has the exclusive right to use the subsoil (within the boundaries of a certain mining field; Article 7). The use of individual areas of subsoil may be restricted or prohibited for reasons of national security and environmental protection (Article 8). The use of subsoil within populated areas, suburban zones, industrial, transportation and communication facilities can be partially or fully prohibited if such use might endanger people's life and health, cause damage to production facilities or the local environment. The use

---

[1] According to: "Federal Law on procedures for Foreign Investments in Companies of Strategic Significance for National Defense and Security (2008)."

of subsoil within specially protected areas must be performed under the appropriate regulations for areas of that status.

Article 9 covers the subsoil users (i. e. operators). The users of subsoil resources are persons engaging in entrepreneurial activities, including foreign citizens and legal entities. Under the terms of production sharing agreements users of subsoil resources can be legal entities as well as associations of legal entities' which are set up on the basis of joint operations agreements (contracts of simple partnership).

Article 10 ML regulates the licensing process. The permit can be granted on basis of a public tender or in the direct application way.

The right to use *subsoil plots* is granted either for a fixed or unlimited period of time. The right to use subsoil plots is granted for a fixed period of time in the following cases (Article 10 ML):

- for the geological study for a period of up to 5 years or for a period of up to 10 years during the works of the geological exploration of the subsoil section in the inland sea waters, the territorial sea and the continental shelf of the Russian Federation;
- in order to extract mineral resources – for the period of extraction of the mineral deposit, which is calculated on the basis of a feasibility study for the extraction of the mineral deposit ensuring the rational use and protection of subsoil.

The issuance of a permit implies the agreement of the land owner (Article 11 ML). Articles 13 and 14 ML regulate the procedure and rejection of applications.

Tenders and Auctions for the Right to Use Subsoil Plots (Article 13 ML)

The adoption of decisions on holding tenders or auctions for the right to use subsoil, must be carried out by the federal body in charge of managing the state mineral reserve, or by the Government of the Russian Federation, when dealing with the continental shelf of the Russian Federation. A decision on the endorsement of the outcome of a tender or auction for the right to use a subsoil plot must be taken within 30 days at most of the date of holding the tender or auction.

The basic criteria for determining the winner when holding a tender for the right to use a subsoil plot is the

- scientific and technical level of programs aimed at geological survey and use of subsoil plots,

- completeness of mineral extraction,
- contribution to the social and economic development of the territory,
- time periods for implementation of appropriate programs,
- efficiency of measures aimed at the protection of subsoil and the environment, due regard to the national security interests of the Russian Federation.

A further criterion for determining the winner when holding an auction for the right to use a subsoil plot will be the amount of the one-time payment for the right to use the subsoil plot.

Rights and obligations of the subsoil user (Article 22 ML)

The subsoil user has the right to use the subsoil area granted to it for any type of entrepreneurial in accordance with the purpose provided for by the permit or by the production sharing agreement. He has further the right to use the results of his activity, including produced mineral raw materials in accordance with the permit. The user of subsoil must ensure the following:

- compliance with technological standards (norms, regulations) for operations related to the use of subsoil;
- contribution of reliable data on explored and recoverable reserves;
- safe performance of operations related to the use of subsoil;
- compliance with standards (norms, regulations) approved in the established procedures for conservation of subsoil, protection of the atmosphere, land, forests, water;
- restoration of land areas to a condition suitable for further use.

Rational Use and Conservation of Subsoil

According to Article 23 ML the operator must ensure full geological exploration, rational comprehensive use and conservation of subsoil. He must ensure the maximum recovery of the main reserves and associated minerals. Furthermore the performance of state evaluation and keeping a state record of mineral reserves is required.

32.2.1.3 Authorities

The mainly responsible authority for mining is the Minister for Natural Resources. The main objective of state regulation of the use of subsoil is ensuring replacement of the mineral raw materials base, its rational use and conservation in the interests of the present and future generations of the people of the

Russian Federation (Article 35 ML). State regulation of the subsoil use must be performed through licensing, registering and control. State regulation has amongst other causes the following objectives:

- to ensure the performance of geological exploration of the Russian Federation territory, its continental shelf, the Antarctic and the World Ocean floor;
- to establish quotas for shipments of produced mineral raw materials;
- to establish standards (norms, regulations) for geological exploration, use and conservation of subsoil;
- to levy payments related to the use of subsoil, as well as control prices of individual types of mineral raw materials.

State control over exploration, rational use and conservation of subsoil must be performed by the state mining inspection bodies jointly with ecological and other controlling bodies. The powers of the state controlling bodies for geological operations, their rights, obligations, and operation rules are defined by statutes subject to approval by the Russian Federation government (Article 37 ML).

32.2.1.4 Fees and Taxes

Article 39 ML regulates the subsoil use payment system. The following payments must be made in the case of use of subsoil:

1. one-off payments for the use of subsoil upon the onset of certain events stipulated by the permit;
2. regular payments for the use of subsoil;
3. payment for geological information on subsoil;
4. a fee charged for participation in a tender (auction);
5. a fee charged for the issuance of licences.

Apart from these, subsoil users shall pay the other taxes and fees established under the Russian legislation on taxes and fees. Subsoil users acting as party to a production sharing agreement shall be deemed payers of subsoil use payments under Russian law.

Furthermore payments for the use of the subsoil are required as follows (Article 43 ML):

1. Regular subsoil use payments will be collected for providing subsoil users with exclusive rights for prospecting and assessment of mineral deposits/fields, mineral resource exploration, and operation of structures not relating to extraction of mineral resources.

2. The rate of regular subsoil use payments will be set depending on the economic and geographic conditions, the size of the subsoil tract, the type of mineral resource, the duration of works, the degree to which the territory has been studied in terms of geology and the degree of risk involved. Regular subsoil use payments will be collected for the area of the permit tract allotted to a subsoil user less the portion of the permit tract that has been returned.

The specific rate of a regular subsoil use payment is set by the executive body of a subject of the Russian region at the proposal of the territorial body responsible for state subsoil management separately for each subsoil tract, for which a subsoil use permit is issued in the established manner and which is located on the territory of an appropriate subject of the Russian Federation, within the following limits (roubles per km² of subsoil tract):

Table 124: Rates of Regular Payments for the Use of Subsoil (Article 43 ML) – Russian Federation

	Rate	
	Minimum	Maximum
1. Rates of Regular Payments for the Use of Subsoil with the Aim of Search and Evaluation of Mineral Deposits		
Precious metals	90	270
Metallic minerals	50	150
Placer deposits of minerals of all types	45	135
Non-metallic minerals, coal, oil shale and peat	27	90
Other hard minerals	20	50
Underground waters	30	90
2. Rates of Regular Payments for the Use of Subsoil with the Aim of Prospecting Minerals		
Precious metals	3.000	18.000
Metallic minerals	1.900	10.500
Placer deposits of minerals of all types	1.500	12.000
Non-metallic minerals	1.500	7.000
Other hard minerals	1.000	10.000
Underground waters	800	1.650

32.2.2 Additional Legal Basics

Table 125: List of Acts significant for Raw Materials – Russian Federation

Name of Law	No. of Law and Year of Issuing
Regarding Validation of the Bases of State Policy in the Sphere of Utilization of Mineral Raw Materials and the Use of Subsoil	Governmental Ordinance No. 494-r of 2003
Validating the Regulation on Forest Management for Exploration of Subsoil and Management of Mineral Deposits (Minister of Natural Resources)	Order No. 109 of 2007
Validating Authorization for the Management of Public Forestland and Transfer thereof in the Category of Non-forest Land (Federal Ecological, Technological and Nuclear Supervision Service)	Order No. 595 of 2006
Validating the Regulation on State Supervision over Exploration and Rational Management of Subsoil	Ministerial Decree No. 293 of 2005
Validating the Regulation on Examination of Applications for Concession of the Plots of Subsoil for Exploration (Ministry of Natural Resources)	Order No. 61 of 2005
Validating the Regulation on Examination of Applications for Exploration of the Plots of Subsoil of Internal Sea and Continental Shelf (Ministry of Natural Resources)	Order No. 62 of 2005
Validating the Regulation on Federal Agency on Subsoil Management	Ministerial Decree No. 293 of 2004
Regarding the Sphere of Competence of the Federal Technological Supervision	Ministerial Decree No. 180 of 2004
Validating the Regulation on Renewal of Permits for Subsoil management (Minister of Natural Resources)	Order No. 1026 of 2003
Regarding Validation of the Regulation on Federal Mining and Industrial Supervision	Ministerial Decree No. 841 of 2001

Name of Law	No. of Law and Year of Issuing
On Environmental Protection	Federal Law No. 7-FZ of 2002, amended by Federal Law No. 93-FZ of 2008
On the Protection of the Atmospheric Air	Federal Law No. 96-FZ of 1999
On Emissions of Pollutants to the Atmospheric Air	Ministerial Decree No. 183 of 2007
Land Code	Law No. 136-FZ of 2001
Federal Law on Land Reclamation	Law No. 4-FZ
Water Code	Law No. 74-FZ of 2007
Implementing Water Code	Federal Law No. 73-FZ of 2006
On Industrial and Consumer Waste	Federal Law No. 89-FZ of 1998
On the Continental Shelf of the Russian Federation	Federal Law No. 187-FZ of 1995
Federal Law on Production Sharing Agreements	1995
Federal Law on Procedures for Foreign Investments in Companies of Strategic Significance for National Defence and Security 57-FZ	2008

33 Serbia

33.1 General Facts

The national territory of Serbia is 77,474 km² and the inhabitants are 7,4 million. The population density is 96/km². The GDP per capita in 2009 amounted to about $ 5.808.

Constitutional Structure

The form of government is a republic. Serbia is divided into 22 districts; these in turn are divided into 162 municipalities.

33.2 Production of Raw Materials

Table 126: Production Date – Serbia (Montenegro included) (World Mining Data, Weber and Zsak, 2008)

Resources		2002	2003	2004	2005	2006	Change 02/06	Change 05/06
Iron	(t)		74.000	73.000	70.000	69.000		−1,43
Aluminium	(t)	0	116.700	120.800	120.400	121.800		1,16
Bauxite	(t)	0	540.100	486.000	610.000	657.300		7,75
Copper	(t)	0	15.500	13.800	11.600	11.100		−4,31
Lead	(t)	0	1.500	900	2.000	7.000		250,00
Zinc	(t)	0	5.600	1.500	2.000	8.000		300,00
Gold	(kg)	0	363	328	330	330		0,00
Platinum	(kg)	0	5	5	5	0		−100,00
Palladium	(kg)	0	20	20	20	0		−100,00
Silver	(kg)	0	2.000	2.400	2.400	2.400		0,00
Asbestos	(t)	0	730	700	620	500		−19,35
Feldspar	(t)	0	4.900	4.500	3.000	4.000		33,33
Gypsum	(t)	0	36.000	3.500	0	50.000		
Kaolin	(t)	0	60.000	58.000	50.000	60.000		20,00
Magnesite	(t)	0	73.000	70.000	60.000	50.000		−16,67

Resources		2002	2003	2004	2005	2006	Change 02/06	Change 05/06
Salt	(t)	0	12.000	11.000	10.500	10.000		−4,76
Sulphur	(t)	0	76.000	75.000	77.000	79.000		2,60
Steam Coal *)	(t)	0	90.000	85.000	80.000	141.000		76,25
Brown Coal	(t)	0	42.000.000	41.000.000	38.700.000	38.907.000		0,53
Natural Gas	(Mm³)	0	700	700	680	670		−1,47
Oil	(t)	0	830.000	820.000	820.000	600.000		−26,83

*) incl. Anthracite

Table 127: Production of aggregates – Serbia (Metric tons unless otherwise specified), (USGS, 2008)

		2002	2003	2004	2005	2006
Lime	1000 metric tons	468	402	400	400	400
Sand and gravel, excluding glass sand	1.000 m³	2.074	1.507	1.500	7.556	8.633
Silica:						
Quartz sand		258 801	260 880	260 000	260 000	260 000
Stone, excluding quartz and quartzite, dimension, crude:						
Ornamental	m²	103 000	69 000	70 000	70 000	70 000
Crushed and broken	1.000 m³	3.000	2.000	2.000	2.000	2.000
Other, stone blocks	m³	1.000	500	500	500	500

33.3 Normative Basics

33.3.1 Primary Legal Basics

The primary legal basics of mineral extraction activity is the Law on Geological Exploration No. 44 of 1995 and the Mining Law of 1995 as amended by Law No. 5 of 2006.

33.3.1.1 General Rules

Table 128: Structure of Law on Geological Exploration – Serbia

Chapter 1	Basic Provisions Articles 1–11
Chapter 2	Execution of Geological Explorations Articles 12–39:
	Conditions and Methodology of Execution of Geological Explorations
	Project of Geological Explorations
	Study of Results of Geological Explorations
	Approval of Explorations
	Exploratory Area
	Other Regulations on Execution of Geological Explorations
Chapter 3	Financing of Programs of Basic Geological Explorations Articles 40–44
Chapter 4	Inspection Articles 45–49
Chapter 5	Authorization for Passing By-Laws Article 50
Chapter 6	Penal Provisions Articles 51–52
Chapter 7	Transitional and Final Provisions Articles 53–59

Table 129: Structure of the Mining Law – Serbia

I Basic Provisions Articles 1–12	
I A Mining Agency Article 12A	
II Mining Of Mineral Raw Materials Articles 13–16A	Mining Approval Articles 17–23
	Mining Field Articles 24–25
	Execution of Mining Operations Articles 27–43
	Other Provisions on Mineral Raw Material Mining Article 44–51
	Cadastre of Mining Fields

III Mining Measurements and Mining Plans Articles 54–59	
IV Protection Measures Articles 60–73	
V Qualifications for the Performance of Certain Activities During The Mining of Mineral Raw Materials Articles 74–81	
VI Inspection Supervision Articles 82–88	
VII Penalty Provisions Articles 89–97	Criminal Offences Corporate Offences Offences
VIII Final and Transitional Provisions Articles 98–101	

Mineral raw materials are classified as follows (Article 3 ML):

1) all types of coal and oil shale;
2) all hydrocarbons in liquid and gaseous state (oil and gas) and other natural gases;
3) radioactive raw materials;
4) all metallic raw materials;
5) raw materials created by mining and processing of mineral raw materials;
6) non-metallic raw material and raw material used for the production of the construction material;

Ownership of mineral rights

Mineral raw materials, as a publicly owned natural resource, can be used under conditions and in the manner stipulated by this law (Article 2 ML).

33.3.1.2 Issuing of Permits

Exploration

The Exploration Law (EL) distinguishes between basic and specific exploration. The basic explorations encompass: explorations of development, composition and structure of the earth crust; establishment of the potential of the region regarding the availability of finding mineral deposits, as well as drawing of appropriate geological maps (Article 5 EL). The basic geological explorations are duties of general interest for the Republic and are financed from the budget of the Republic of Serbia. The exploration area, i. e. terrain on which adequate basic geological explorations have not been performed, are not subject of detailed geological explorations of mineral deposits (Article 7 EL).

Monitoring of explorations will be conducted by the ministry in charge of geology (Article 11 EL).

Exploration can be undertaken by an operator or another legal entity registered in the court registries for execution of these works (Article 12 EL). Use of data and documentation on basic geological explorations are subject to payment of a fee up to 5 % of the actual value of the executed exploration in the exploratory area (Article 15 EL). Exploration must be executed pursuant to the project which should specifically include: terms of reference for the project; general data on exploratory area; overview of results of earlier explorations; schedule of execution of works and timeframes for completion thereof (Article 17 EL).

Articles 21–23 EL include detailed provisions related to the exploration conduction. The exploration permit will be issued by the competent authority. According to Article 21 EL the operator must execute the works in accordance with the project of geological exploration, technical regulations, norms, standards and modern scientific and expert methods applied in the execution of these works and observe all the stipulated measures of safety at work.

In the course of execution of the geological exploration, the manager of exploration should ensure expert supervision for the execution of geological exploration. The expert supervision includes: control of the quality of exploratory works executed and fulfilment of regulations, standards, technical and quality norms; verification of conformity of exploratory works with the project of geological explorations; control of the application of measures of safety at work and environmental protection; monitoring the schedule of execution of works and investments; observe contract timeframes (Article 22 EL).

The manager of explorations may perform expert supervision on the execution of geological explorations directly or cede such supervision to a company or another legal entity specialized in the elaboration of projects for the particular type of exploratory works (Article 23 EL).

Exploitation

Article 13 ML explicitly regulates the extraction of raw materials. Mining of raw materials can be carried out by natural or legal persons registered for the performance of this activity and licensed for the performance of mining operations (Article 13 ML).

Article 17 ML regulates the procedures for the mining approval. Mining approval is issued upon the application of the operator in the area where mining will be performed (Article 17 ML). Mining is carried out on the basis of the mining approval, within the time period specified by the approval and according to the schedule and in the amount anticipated by the mining project. The following documents according to Article 18 ML must be submitted together with the application for the issuing of the mining approval:

1) A layout map, in the scale of 1:25 000, together with public transport routes;
2) A certificate on balance reserves of raw materials issued on the basis of investigations;
3) A feasibility study on mining of deposits with an overview of conditions of mining, environment protection measures and social impacts, necessary funding to be engaged;
4) A land reclamation project.

Also agreements with additional authorities are necessary:

The authority responsible for zoning activities in terms of alignment of mining with corresponding spatial, i.e. zoning plans; the ministry of agriculture and forestry, when mining is performed on agricultural and forest land; the ministry of environment; the ministry of water resources management, if water table is influenced by mining.

Additionally information should be provided concerning the type of mineral raw materials to be mined; the production capacity according to balance reserves and the position and exact boundaries of the mining field.

Article 21 ML determines the mining area. The operator which was given the mining approval for certain raw materials, can perform at the same mining field the mining of other mineral raw materials which were not included in the given approval.

Based on the data on identified conditions and quality of raw materials and mining conditions, the operator must develop a long-term mining program for the period of at least of 10 years and annual operational plans for the execution of mining operations. The enterprise must adopt annual plans at the latest November 30 of the ongoing year for the forthcoming year (Article 27 ML).

Approval for the execution of mining operations

The setting up of mining plans requires an approval issued by the Ministry of Energy and Mining (Article 35–38 ML). The operator must perform land reclamation in accordance with the land reclamation plan, during and upon finalisation of mining operations, at the latest within one year from the day of the completion of operations and in areas where mining operations were completed. This implied that land protection measures for the purpose of securing people's lives, health and property should be undertaken. Ministry competent for agricultural, water resources management, i.e. environmental activities must be informed (Article 48 ML).

The following documents or permits must be submitted together with the application for the issuing of the approval for the execution of mining operations (Article 36 ML):

1) The mining report with the certificate of the enterprise;
2) A certificate on certified balance reserves, issued in accordance with the regulations on geological investigations;
3) Evidence on the proprietary or utilization right, i. e. servitude right on the land specified for the mining;
4) A compliance of the competent authority for environmental activities;
5) Permit of water authority, if execution of mining operation has an impact on the water table;
6) A reclamation report.

33.3.1.3 Authorities

The main responsible authority for mining is the Ministry of Energy and Mining. The ministry has its own inspectors. The roles of inspectors are regulated by Articles 47–48 ML.

Inspection supervision over the enforcement of the provisions of the Mining Law and regulations adopted for its enforcement is performed by the Ministry through its republic mining inspectors (Article 82 ML). The mining inspector is authorised to verify whether: 1) prescribed safety measures and norms are applied; 2) mining is performed on the basis of the mining approval, and in compliance with the mining contract and the approval for the execution of mining operations (Article 85 ML).

According to Article 94 ML the operator (i. e. "legal person") will be fined with 100.000 to 2.000.000 dinars for the corporate offence, if he does not provide optimal technical and economic recovery of raw material deposit in accordance with modern scientific achievement as well as protection measures for the safety of people and property (Article 9). According to Article 95 ML the operator will be fined with 50.000 to 500.000 dinars for an offence, if he does not submit the annual operations plan for the opening and mining of mineral raw materials (Article 27, paragraph 2) within the specified deadline.

33.3.1.4 Fees and Taxes

The operator performing the mining of mineral raw materials has to pay a royalty for the utilisation of raw materials. Royalty is set in accordance with the following bases (Article 16 ML):

- for all metallic raw materials 3 %;
- for raw materials created by mining and processing 4 %;
- for non-metallic raw material and raw materials used for the production of the construction material 5 %;
- for all types of salts and salt waters 1 %.

33.3.2 Additional Legal Basics

Table 130: List of Acts significant for Raw Materials – Serbia

Name of Law	No. of Law and Year of Issuing
Law on Environmental Protection	Law No. 135 of 2004
Law on Environmental Impact Assessment	Law No. 135 of 2004
Law on Foreign Investment	2002
Law on General Administrative Procedure	
Official Statistics Law	
Civil Procedure Code	
Foreign Investment Law	
Free Economic Zones Law	
Law on Prices	
Law on Business Companies	
Law on Serbian Business Registers Agency	
Law on the Registration of Business Entities	
The Law on Determination and Classification of Mineral Raw Materials Reserves and Geological Exploration Data Presentation	
Law on Tax Procedure and Tax Administration	
Property Tax Law	
Law on Planning and Construction	

Law on environmental protection

This Law regulates the impact assessment procedure for projects that may have significant effects on the environment, the contents of the Environmental Impact Assessment (EIA) Study, the participation of authorities and organisations concerned, the public participation (Article 1). According to Article 3 mining is subject to EIA.

Environmental Impact Assessment Law

The Environmental Impact Assessment Law regulates the impact assessment procedure for projects that may have significant effects on the environment, the contents of the Environmental Impact Assessment Study, the participation of authorities and organisations concerned, the public participation, transboundary exchange of information for projects that may have significant impact on the environment of another state, supervision and other issues of relevance to impact assessment (Article 1).

34 Slovakia

34.1 General Facts

The national territory covers 49,035 km² and the inhabitants are 5.379.455. The population density is 111/km². The GDP per capita for 2009 is estimated at $ 16.281.

Constitutional Structure

The form of government is a parliamentary republic. It is divided into 8 self-governing regions resp. counties (krajov), each named after the principal city.

34.2 Production of Raw Materials

Table 131: Production Data – Slovakia (World Mining Data, Weber and Zsak, 2008)

Resources		2002	2003	2004	2005	2006	Change 02/06	Change 05/06
Iron	(t)	174.570	200.000	385.000	258.500	240.000	37,48	–7,16
Aluminium	(t)	111.600	131.400	156.900	159.200	157.300	40,95	–1,19
Copper	(t)	100	95	93	65	0	–100,00	–100,00
Gold	(kg)	78	79	107	109	100	28,21	–8,26
Silver	(kg)	95	90	70	60	0	–100,00	–100,00
Baryte	(t)	25.820	14.000	5.600	4.200	4.100	–84,12	–2,38
Bentonite	(t)	90.000	120.000	193.291	120.000	100.000	11,11	–16,67
Diatomite	(t)	950	940	930	910	800	–15,79	–12,09
Gypsum	(t)	121.700	115.000	127.100	107.500	120.000	–1,40	11,63
Kaolin	(t)	24.600	40.000	55.385	27.730	25.000	1,63	–9,84
Magnesite	(t)	1.464.500	1.640.900	965.900	1.555.000	1.600.000	9,25	2,89
Perlite	(t)	18.630	30.000	42.400	99.900	99.000	431,40	–0,90
Salt	(t)	102.700	104.800	104.300	105.100	106.000	3,21	0,86
Talc	(t)	2.290	4.000	7.100	200	300	–86,90	50,00

Resources		2002	2003	2004	2005	2006	Change 02/06	Change 05/06
Lignite	(t)	3.661.280	3.508.820	3.101.790	2.513.000	2.201.000	−39,88	−12,42
Nat. Gas Mio	(Mm³)	200	187	178	151	150	−25,00	−0,66
Oil	(t)	51.770	47.943	42.082	33.150	32.000	−38,19	−3,47

Table 132: Aggregates Production – Annual Statistics/Slovakia, 22. April 2008 (UEPG 2008)

Sand & Gravel (1)		Crushed rock (2)		Marine Aggregates		Recycled Aggregates (3)		Manufactured Aggregates (4)	
2004	2006	2004	2006	2005	2006	2005	2006	2005	2006
8,9	10,0	16,9	16,5	n. a.	0,0	0,2	0,2	0,3	0,3

(1) Sand and Gravel: sold production including crushed gravel
(2) Crushed rock: sold production (excluding crushed gravel)
(3) Recycled Aggregates: materials coming from construction and demolition waste used in aggregates market
(4) Manufactured aggregates include blast-furnace-slag, electric-arc-furnace-slag, incinerator bottom ash (IBA), pulverised fuel ash (PFA)

34.3 Normative Basics

34.3.1 Primary Legal Basics

The primary legal basics of mineral extraction activity is the Mining Law ('Mining Act') No. 44 of 1988 as amended by Law No. 114 of 2010.

34.3.1.1 General Rules

Table 133: Structure of the Mining Law

Part I Basic Provisions	Article 1 Introductory Provisions	
	Article 2 Minerals	1–2
	Article 3 Division of minerals into reserved and non-reserved ones	1–3
	Article 4 Mineral deposit	
	Article 5 Mineral wealth	1–2

	Article 6 Exclusive deposit	1–2
	Article 7 Deposit of non-reserved minerals	
	Article 7a Organisation	
	Article 8 Repealed	
	Article 9 Repealed	
	Article 10 Duties of the organisation in utilisation of the exclusive deposit	1–2
Part II Exploration of Deposits and Management of Reserves of Exclusive Deposits	Article 11 Prospecting and exploration of exclusive deposits	
	Article 12 Report on the occurrence of the reserved mineral deposit	1–3
	Article 13 Reserves of the exclusive deposit and conditions of their use	1–3
	Article 14 Classification of exclusive deposit reserves, assessment and approval of the exclusive deposit reserves estimation	
	Article 14a Writing off the exclusive deposit reserves	
Part III Protection of Mineral Wealth	Article 15 Assurance of mineral wealth protection in territorial arrangements	1–2
	Article 16 Deposit reservation	1–3
	Article 17 Determination of the deposit reservation	1–8
	Article 18 Limitation of some activities in the deposit reservation	1–2
	Article 19 Permit for buildings and facilities in the deposit reservation	1–2
	Article 20 Exclusive deposit reserves management	1–2
	Article 21 Writing off the reserves of exclusive deposits	
	Article 22 Recording of the exclusive deposit reserves	1–4
Part IV Construction of Mines and Quarries	Article 23 Designing, construction and reconstruction of mines and quarries	
Part V Mining of exclusive Deposit	Article 24 Power to mining the exclusive deposit	1–6
	Article 25 Mining reservation	1–3
	Article 26 Boundaries of the mining reservation	1–3
	Article 27 Determination, changes and cancellation of the mining reservation	1–9

	Article 28 Proceedings in determination, changes and cancellation of the mining reservation	
	Article 29 Keeping records	1–6
	Article 30 Rational utilisation of exclusive deposits	
	Article 31 Duties and rights of the organisation in mining of the exclusive deposit	1–3
	Article 32 Plans of the development and mining of the exclusive deposits and plans of security and liquidation of the main workings and quarries	1–5
	Article 32a Reimbursements	1–10
	Article 33 Solution of conflicts of interests	1–7
Part VI Other Interventions into the Earth's Crust	Article 34 Special interventions into the earth's crust	
	Article 35 Old workings	1–5
Part VII Mining Damages	Article 36 Mining related damages	1–3
	Article 37 Liquidated mining damages	1–7
Part VIII Common Provisions	Article 38 Safety of operation	
	Article 39 Mine surveying and geological documentation	1–3
	Article 40 Mine waters	
	Article 41 Relation to the Administrative Proceedings	
	Article 42 Repealed	
Part IX Transient and Final Provisions	Article 43 Transient provisions	1–5
	Article 44 Repealing provisions	1–2
	Article 45 Effectivity	
Appendix I	List of accepted legal Acts of the European Communities and European Union	1–5

Minerals classification

Article 3 ML classifies minerals into reserved and non-reserved ones

The reserved minerals are

a. radioactive minerals,
b. all types of coal, oil and combustible natural gas and bituminous rocks,

c. minerals, from which it is possible to obtain metals in the industrial manner,
d. magnesite,
e. minerals, from which it is possible to obtain phosphorus, sulphur and fluorine or their compounds in the industrial manner,
f. halite, potassium, boron, bromine and iodine salts,
g. graphite, baryte, asbestos, mica, talc, diatomite, glass sand and foundry sand, mineral pigments, bentonite,
h. minerals, from which it is possible to obtain rare earth elements and elements with properties of semi-conductors in the industrial manner,
i. granite, granodiorite, diorite, gabbro, diabase, serpentine, dolomite and limestone, until they are workable in blocks and polishable and travertine,
j. technically usable crystals of minerals and gemstones,
k. halloysite, kaolin, ceramic and refractory clays and claystones, gypsum, anhydrite, feldspars, perlite and zeolite,
l. quartz, quartzite, limestone, dolomite, marl, basalt, phonolite, trachyte, unless these minerals are suitable for chemical-technological processing by melting,
m. mineralised waters, from which reserved minerals can be obtained in the industrial manner.

The other minerals are non-reserved ones.

Ownership of mineral rights

Minerals ("mineral wealth") are owned by the Slovak Republic. According to Article 5 ML the deposits of reserved minerals ("exclusive deposits")[1] only form mineral wealth according to this act.

Acquiring minerals

The operator, for which the area for prospecting and exploration of the exclusive deposit has been determined, has the right of priority as to determination of the mining area (Article 24 ML).

---

1 Regarding the term "reserved mineral": According to Article 6 ML: If a reserved mineral is ascertained in the quantity and quality allowing to expect reasonably its accumulation, the Slovak Geological Institute will issue the certificate of the exclusive deposit.

34.3.1.2 Issuing of Permits

Exploration

According to Article 11 ML the operator must explore the exclusive deposit in such a way as to identify and assess all usable minerals. Furthermore, he must verify the development and position of the exclusive deposit, in order the construction of mines and quarries and the development, preparation and mining of the exclusive deposit could be designed and implemented according to the principles of the mining technology and in order to assure rational utilisation of the reserves of the exclusive deposit.

Report on the occurrence of the reserved mineral deposit

Any person, who has discovered the occurrence of a reserved mineral deposit, outside the organised prospecting, must immediately notify the Slovak Geological Institute. After verification of the occurrence of the reserved mineral deposit, the Slovak Geological Institute issues the certificate to its finder and remunerates him. The amount of this remuneration will be determined considering the scientific, technical and economical significance of such a finding (Article 12 ML).

Exploitation

Power of operator to mine the exclusive deposit is established by determination of the mining reservation (Article 24 ML). The operator may mine the exclusive deposit in the determined mining area after it was granted the permit for mining activity issued by the Subdistrict Mining Office. The operator has the right of handling the mined minerals in the extent and under the conditions specified in the decision on determination of the mining reservation.

The boundaries of the mining reservation on the surface are determined by the closed geometrical figure with straight sides, whose apices are determined by the co-ordinates given in the valid co-ordinate system. The mining area can also be determined at depth (Article 26 ML).

Duties and rights of the operator

If during mining, the deposit of another reserved mineral is discovered in the mining reservation, the operator must notify the Slovak Geological Institute and the Subdistrict Mining Office. If it has been verified by exploration that the discovered deposit can be mined and its mining by other operators would not be rational, the Subdistrict Mining Office can order the operator to mine this exclusive deposit (Article 31).

In order to mine the exclusive deposit to establish buildings and facilities within the boundaries of the mining area, and if it is necessary, also outside these boundaries, which are necessary for the development and mining of the exclusive deposit and for dressing and beneficiation of minerals performed in relation with their mining and for transport of all necessary equipments and materials.

According to Article 32 ML, mining plans of the exclusive deposits are required. These plans shall assure a sufficient start of the exclusive deposit development using suitable mining methods and safety operation. Before stopping operation, the operator must prepare security plans. Details of mining plans and security plans will be stipulated in the legal regulation issued by the Slovak Mining Agency.

34.3.1.3 Authorities

The main responsible authority for mining is the Minister of Economics. Regarding the determination of the deposit reservation the Subdistrict Mining Office (Minister of Economics) is the competent authority (Article 17). The Subdistrict Mining Office (with other relevant authorities) also must create the conditions for protection and rational utilisation of mineral extraction. It further must control the mining activities (Article 20).

34.3.1.4 Fees and Taxes

The Mining Act prescribes to the operators in case of the reserved mineral deposit exploitation to pay for the Mining Area as well as for the exploited minerals. The payment for the Mining Area is paid in the tariff of € 663,88 per every started km² per year.[2]

The payment for exploited minerals is calculated as a percentage of the price of the sold minerals or products made from sold minerals depending on the type of mineral. The fee ranges from 0,1 % to 10 %.

In the case that payment for the Mining Area or exploited minerals is not realised in the specified time, the Local Mining Office claims the payment in accordance with the Administration Fee Act.

---

[2] Bauer, V. (2004), Country Report Slovakia, in: Department of Mining and Tunnelling, lc.

34.3.2 Additional Legal Basics

Table 134: List of Acts significant for raw materials – Slovakia.

Name of Law	No. of Law and Year of Issuing
Government Decree about the Non-Reserved Mineral Deposits Utilisation	Law No. 520 of 1992
Building Act	Law No. 50 of 1976
Act about Mining Activity, Explosives and the State Mining Authorities	Law No. 51 of 1988, as amended by Law No. 154 of 1995
Nature and Country Protection Act	Law No. 222 of 1996, as amended by Law No. of No. 543 of 2003
Environmental Impact Assessment Act	Law No. 127 of 1994
Forestry Act	Law No. 61 of 1978
Agricultural Soil Fund Protection Act	Law No. 307 of 1992
Air Protection Act	Law No. 309 of 1991
Water Act	Law No. 138 of 1975
Land Register Act	Law No. 162 of 1996
Environmental Act	Law No. 17 of 1996
Law on Waste Management	Law No. 238 of 1991, as amended by Law No. 255 of 1993
Act Coll. on Administrative Procedures as amended	Law No. 71 of 1967
The Cadastre Act	1995
Act on Protection of Competition	Law No. 136 of 2001
State Monuments Conservation Act	Law No. 27 of 1988
Commercial Code	Law No. 513 of 1992

Nature and Country Protection Act

The purpose of the Nature and Country Protection Act is to contribute to the conservation of diversity of conditions and forms of life on the earth by developing conditions for the permanent survival, recovery and rational exploitation of natural resources, preservation of natural heritage and the characteristic appearance of the landscape and the achievement and conservation of ecological stability (Article 1).

35 Slovenia

35.1 General Facts

The national territory covers 20,273 km² and the inhabitants are 2.053.355. The population density is 99,6/km². The GDP per capita for 2009 is estimated at $ 23.744.

Constitutional Structure

The form of government is a parliamentary republic. It is divided into 4 administrative divisions, since 2006 it is divided into 210 municipalities.

35.2 Production of Raw Materials

Table 135: Production Data – Slovenia (World Mining Data, Weber and Zsak, 2008)

Resources		2002	2003	2004	2005	2006	Change 02/06	Change 05/06
Aluminium	(t)	87.600	118.305	120.700	138.000	139.600	59,36	1,16
Bentonite	(t)	201	52	100	140	130	−35,32	−7,14
Salt	(t)	138	4.273	5.000	4.200	1.624	1076,81	−61,33
Lignite	(t)	4.684.709	4.828.499	4.800.000	4.539.556	4.520.754	−3,50	−0,41
Nat. Gas	(Mm³)	57	49	217	44	4	−92,98	−90,91
Oil	(t)	613	526	490	303	284	−53,67	−6,27

Table 136: Production of aggregates – Slovenia (Metric tons unless otherwise specified), (USGS, 2008)

		2003	2004	2005	2006
Sand and gravel, excluding glass sand	1.000 metric tons	11.012	12.373	12.432	17.873
Stone, excluding quartz and quartzite, crude:					
Dimension		12.603	10.667	15.262	15.000
Crushed		10.000.000	10.000.000	12.176.491	7.547.058

35.3 Normative Basics

35.3.1 Primary Legal Basics

The primary legal basics of mineral extraction activity is the Mining Law ("Mining Act") No. 56 of 1999 as amended by Law 68 of 2008. (A new Mining Act was published 2010, which will become applicable in 2011)

35.3.1.1 General Rules

Table 137: Structure of the Mining Law – Slovenia

I General Provisions	Article 1 Contents and Purpose of the Act
	Article 2 Definitions
	Article 3 Mineral Resources
	Article 4 Mining Operations
II Programming and Planning of Mineral Resource Management	Article 5 National Mineral Resource Management Programme
	Article 6 General Plan
	Article 7 Individual Mineral Resource Management Plan
	Article 8 Adoption
	Article 9 Expert Tasks
	Article 10 Relation to Spatial Planning

III Exploration and Exploitation of Mineral Resources	Article 11 Exploration and Exploitation
	Article 12 Exploration and Exploitation Areas
	Article 13 Mining Right and Preliminary Exploration Permit
	Article 14 Compensation for Mining Right
	Article 15 Act for Granting a Mining Right
	Article 16 Granting of Mining Rights
	Article 17 Application on the Interest in Granting the Mining Rights
	Article 18 Restrictions on the Granting of Mining Rights
	Article 19 Concession Contract
	Article 20 Priority/Pre-emptive Right
	Article 21 Transfer of the Mining Right
	Article 22 Right to the Same Area
	Article 23 Manner of Concession Relationship Termination
	Article 24 Termination of Concession Contract
	Article 25 Revocation of the Mining Right
	Article 26 Other Matters Concerning Mining Rights
	Article 27 Acquisition of Mineral Resources in the Process of Construction Works

	Article 28 Reporting
	Article 29 Study of Reserves and Sources
	Article 30 Reserves and Sources of Mineral Resources
	Article 31 Business Secret/Secrecy
IV Restrictions on Land related Proprietary Rights	Article 32 Definition
	Article 33 Proprietary Relations
	Article 34 Expropriation Beneficiary
	Article 35 Public Interest
	Article 36 Right of Way
	Article 37 Annulment of Expropriation Order
V The Mining Fund of the Republic of Slovenia	Article 38 Mining Fund
VI Performing of Mining Operations	Article 39 Conditions for Performing Mining Operations
	Article 40 Manner of Performing Mining Operations
	Article 41 Several Operators of Mining Operations
	Article 42 Technical Regulations and Ensuring Health and Safety in Mining Operations
	Article 43 Rescue Service, Notification and Conducting Rescue Operations

	Article 44 Performing of Expert Tasks Related to Safety and Health at Work
	Article 45 Technical Management and Supervision
	Article 46 Authorisations
	Article 47 Mine Supervision Records
	Article 48 Types of Permits and Competence to Issue Permits
	Article 49 Exploration Permit, Exploitation Permit and Permit to Cease Exploitation
	Article 50 Operating Permit
	Article 51 Standardised Permit
	Article 52 Temporary Operating Permit
	Article 53 Final Permit
	Article 54 Commencement of Operations
	Article 55 Permit to Use
	Article 56 Qualifications for Supervising Mining Operations
	Article 57 Technical Approval
VII Suspension of Operations and Permanent Abandonment of Mining Operations	Article 58 Temporary Suspension
	Article 59 Total and Permanent Abandoning of Exploitation

	Article 60 Rehabilitation of The Environment and Remedying of Consequences by Land Reclamation
	Article 61 Termination of Rights and Obligations
	Article 62 Costs of Remedying the Consequences of Mining Operations
VIII Technical Documentation and Designing	Article 63 Technical Documentation
	Article 64 Elaboration of Technical Documentation
	Article 65 Mining projects
	Article 66 Elaboration of Mining projects
	Article 67 Person for the Elaboration of Mining projects
	Article 68 Conditions for Performing the Work of Independent Planner
	Article 69 Authorised Person
	Article 70 Statement of Compliance with Article 64
	Article 71 Revisor
	Article 72 Revision Clause
	Article 73 Progress of Revision
	Article 74 Mineral Resources or Rock Structures Exploration Project
	Article 75 Mineral Resources Exploitation or Abandoning of Exploitation Project

	Article 76 Construction and Use of Underground Openings Project
	Article 77 Project for Operation
	Article 78 Deviation from the Project for Operation
	Article 79 Mining Plan
	Article 80 Mineral Resources Exploitation Programme
	Article 81 Geological and Geomechanical Research, and Geological Documentation in Mining
	Article 82 Documentation on the Environmental Impact of Mining Operations
	Article 83 Documentation on the Classification by Hazardous Occurrences
	Article 84 Specialized State Exam
IX Mine Inspection	Article 85 Mine Inspection
	Article 86 Mine Inspector's Qualifications
	Article 87 Mine Inspector
	Article 88 Inspector's Rights
	Article 89 Inspector's Responsibilities
	Article 90 Prodcedural Order
	Article 91 Restrictive Injunction

	Article 92 Order	
	Article 93 Operator's Responsibilities	
	Article 94 Other Orders	
X Physical Planning and Protection	Article 95 Environmental Pressures	
	Article 96 Physical Development Permit for Exploitation Area	
	Article 97 Construction in the Exploitation Area	
	Article 98 Use of Water	
XI The Chamber of Engineers of Slovenia and the Central Section of Mining Engineers and Engineers of Geotechnology	Article 99 Joining into the Engineering Chamber of the Republic of Slovenia and the Foundation of the Central Section	
XII Penal Provisions	Article 100 Offences	
	Article 101 Offences	
	Article 102 Offences	
	Article 103 Mandatory Fine	
	Article 104 Mandatory Fine	
XIII Transitional and Final Provisions	Article 105 Granting of Mining Rights in the Transitional Period	
	Article 106 Transition to a New System of Programming and Planning of Mineral Resources Management	

Article 107	Rehabilitation of Wells and Disused Mining Facilities
Article 108	Adjustment of Tolar Amounts
Article 109	Procedures Initiated before the Entering into Force of the Present Act
Article 110	
Article 111	Regulations Issued by the Minister Responsible for Mining
Article 112	Government Regulations
Article 113	Validity of Documents, Recognition of the Existing Professional Educational Level
Article 114	International Agreement
Article 115	Validity of Existing Regulations
Article 116	Termination of Validity of Acts
Article 117	Date of Validity of the Act

Classification of minerals

Mineral resources are all energy resources (fossil fuels and hydrocarbons), ores and rocks, parts of rocks, minerals, deposits, evaporites (salt) and geothermal energy sources comprising (Article 3 ML):

1. all types of coal,
2. oil and bituminous rocks,
3. radioactive mineral resources,
4. mineral resources from which metals and their usable compounds can be produced
5. all types of natural gas,
6. waters, which mineral substances can be extracted,
7. graphite, sulphur, magnesite, fluorite, barite, mica, gypsum, calcite, chalk, bentonite, chert, quartz, flint and gravel, kaolin, fire-resistant clay, marl and limestone, dolomite for industrial purposes, feldspar, diatomaceous earth, puzzolan tuff, kyanite, leucite, zeolite tuff,

8. natural stones (ornamental architectural stone, architectural construction stone),
9. semi-precious and precious stones,
10. all kinds of salts and salt waters,
11. dimensional stone,
12. gravel, sand, fine sand, clay flysch and marl,
13. potter's clay, ceramic clay and brick clay.

Ownership of mineral rights

Mineral resources are the property of the State. The right to explore and exploit mineral resources ("mining right") and the permit for prospecting is obtainable under the conditions and in the manner provided by the Act and by relevant regulations issued on the basis of this Act (Article 11 ML).

Acquiring mineral rights

The basis for granting a mining right is a concessionary act. A concessionary act is a Government regulation issued in accordance with the National Programme (i. e. National Mineral Resource Management Programme) mentioned in Article 5 ML. A mining right is the right to explore and/or exploit mineral resources for the purposes of industry and trade. The mineral right holder is a legal or natural person who has obtained the mining right by concession in accordance with the provisions hereto. The procedures of marking boundaries, keeping the register of mining right holders, and the methods of keeping a land register of the exploration and working areas are prescribed by the Minister responsible for mining (Article 15 ML).

The State and the mining right holder must regulate their relationship concerning a mining right in a concession contract. The relationship between the State and the mining right holder are terminated: by termination of the concession contract, by revocation of the mining right, by expiry of the mining right (Article 19 ML).

35.3.1.2 Issuing of Permits

Prospecting and exploration

A prospecting permit can be obtained by a legal or a natural person, subject to the permit of the administrative unit of the area where the exploration takes place. This permit is issued for the period of the execution of preliminary exploration programme, but for not more than one year. An applicant must attach to his application: working programme, a layout plan with marked boundaries of the area to be prospected, approval of the communities in the area where prospecting will take place.

An exploration is granted for a period of not more than five years, and can be extended for three years maximum, each time when the scope of exploration works defined by the concession contract has not been completed despite regular and well performed exploration (Article 13 Mining Act). Spatial plans of state and local communities are base for beginning a process of concession procedure. An application for exploration permit has to be accompanied by:

- evidence that he has been granted a mining right for exploration,
- a revised mining project for the exploration,
- a permit for land development activities in the course of exploration;
- a layout plan with marked boundaries of the exploration area on a scale which makes it possible to determine in nature the boundaries of the exploration area, including the description of the exploration area location.

The holder of a mining right for the exploration of mineral resources has at least once a year submit to the Ministry responsible for mining a report on the results of mineral resources exploration.

Exploitation

An exploitation concession is granted for not more than 50 years, except in the case that, because of substantial investment in the exploitation of a certain mineral resource in a certain area, the reserves in the exploitation area cannot be extracted in full. The State must grant a mining right on the basis of a public tender carried out by the Ministry responsible for mining in accordance with Article 15 ML.

Mining operations can be performed by a company or an independent contractor, registered in the Republic of Slovenia for the corresponding activity. Where the holder of a mining right does not perform the mining operations himself, he must make an appropriate contract with an operator who complies with the conditions. The compliance with the prescribed conditions for performing activities is determined by the mine inspector (Article 39 ML).

Mining Plan (Article 79 ML)

For the proper performing of mining operations and rational exploitation of mineral resources, to ensure safe operation and to prevent material damage, the operator must elaborate mining plans and maps based on measurements and showing the condition of mining operations, their relative position, as well as their location with respect to ancient workings, facilities and watercourses on the surface. The operator must deliver copies of the mining plans to the Ministry responsible for mining, at the said Ministry's request. The Minister responsible for mining prescribes the procedure and conditions for performing underground measurements, for the elaboration of mining plans and maps, their scale, the selection of a co-ordinate system, as well as keeping the records of underground measurements and other documentation used in the elaboration of the technical documentation.

Rehabilitation (Article 60 ML)

After obtaining a permit for the termination of mineral resources exploitation referred to Article 48 ML, the mining right holder must perform the final rehabilitation of the environment and land reclamation to remedy the consequences of mining operations. The operator of other mining operations as defined by the present Act is also obliged to rehabilitate the environment and remedy the consequences of mining operations. In areas where total rehabilitation and land reclamation is not possible, the operator is obliged to undertake safety measures to prevent threatening to the life and health of people and animals as well as potential causes of environmental pollution and/or of foreseeable damage to structures and the environment.

35.3.1.3 Authorities

The main responsible authority for mining is the Minister of Environment, Spatial planning and Energy (Energy Office).

The inspection of the implementation of the provisions of the Mining Act has to be performed by the Mine Inspector. The Mine Inspector's office must co-operate in its activities with other inspection offices as well as with expert organisations in mining (Article 85 ML).

Where supervising the implementation of the measures of health and safety in mining operations, an inspector has equal rights as a labour inspector pursuant to the provisions of the act regulating labour inspection and the act regulating health and safety. An inspector is also qualified to supervise the physical development caused by mining operations with the purpose of mineral resources exploration and exploitation after the physical development permit has entered in force, and in the case where an inadmissible environmental intervention has been made (Article 86 ML).

35.3.1.4 Fees and Taxes

All companies must pay royalty for all exploited mineral resources; they also must pay fee for land used for mining operations. They are also paying a pay back fee for remediation of mine sites or they must get a bank guarantee for the sum of remediation works. There is a facilitation in paying royalties and fees in the last two years of mining operation.[1]

For the purpose of implementing the mining right the mining right holder is liable to pay compensation to the state for concession in accordance with the provisions of the Mining Act and related regulations issued on the basis of this Act. The basis for the calculation of such compensation is the average price of a mineral resource unit, and depends on the type, extent and occurrence of the mineral resource (Article 14 ML).

The payment for the implementation of a mining right for exploration must be effected in lump sum when signing the contract and may amount to a maximum of 100.000 tolars per hectare of the exploration area. The payment for the implementation of a mining right for exploitation is payable in annual amounts and may be maximum 100.000 tolars per hectare of exploitation area and maximum 20% of the average price of one a unit of mineral resources produced in the year concerned. The payment shall fall to the State and to the community in equal portions, i. e. 50:50. The local community should spend that money only for determined purposes. The State share is an income of Mining Fund, which can finance or offer loans for exploration of mineral resources, dispatching of consequences of exploration and exploitation of mineral resources, new mining technologies etc. (Article 14 ML).

35.3.2 Additional Legal Basics

Table 138: List of Acts significant for raw materials – Slovenia

Name of Law	No. of Law and Year of Issuance
Spatial Planning Act	2003
Environmental Protection Act	Law No. 32 of 1993
Water Act	Law No. 67 of 2002
Nature Conservation Act	Law No. 56 of 1999
Administrative Fee Act	

---

1 Solar, S. (2004), Country Report Slovenia, in: Department of Mining and Tunnelling, lc.

Name of Law	No. of Law and Year of Issuance
National Statistics Act	Law No. 45–2169 of 1995, last amended by Law No. 9–529 of 2001
Public administration Act	
Civil Code (Slovene)	
Companies Act (ZGD-1)	Law No. 001–03/92–1/69 of 2006
Competition Act (ZPOmK-1)	
Law on Health and Safety at Work Act	Law No. 160–01/95–1/9 of 1999
Real Property Transaction Tax Act	Law No. 434–02/98–18/5 of 2006
Forest Act	1995
Nature Conservation Act	1999
Regulations on Changes and Additions to the Regulations on the Management of Wastes which Contain Toxic Substances	1996
Regulations on Initial Measurement of Noise and Operational Monitoring for Sources of Noise and on Conditions for their Execution	

Environmental Protection Act

The Environmental Protection Act (EPA) is a fundamental act regulating the objectives and principles of environmental protection. The purpose of the legislation is to protect the closely connected living and natural environment and to direct the developmental processes and activities affecting the environment based on the balanced developmental and environmental needs. The EPA provides the principles, the basic instruments and the institutes for the regulation of the legal protection of the environment which form a framework for the preparation of all other acts regulating the specific environmental protection fields.

Water Act

The objective of the management of water is to ensure protection against the adverse effects of waters, to preserve and balance water quantities, and to promote the sustainable use of waters for various types of use, by taking into account the long-term protection of available water sources and their quality (Article 2).

Nature Conservation Act

The Nature Conservation Act (the NCA) is the fundamental regulation in the field of the conservation of biodiversity of wild plant and animal species. The conservation of biodiversity in nature is closely linked to the maintenance of the natural equilibrium. In compliance with the environment conservation development, only those human activities are permitted which meet human needs in a reasonable manner. Pursuant to the NCA, the nature conservation measures and the system for the protection of valuable natural features are taken into account in the spatial planning and in the use and exploitation of natural assets in a way stipulated by the law.

36 Spain

36.1 General Facts

The national territory covers 506.992 km² and the inhabitants are 46.661.950. The population density is 79/km². The GDP per capita for 2009 is estimated at $ 30.862.

Constitutional Structure

Spain is a parliamentary democracy and constitutional monarchy. Political power lies in a central government and 17 autonomous communities. There are also 2 autonomous cities (Ceuta and Melilla).

36.2 Production of Raw Materials

Table 139: Production Data – Spain (World Mining Data, Weber and Zsak, 2008)

Resources		2002	2003	2004	2005	2006	Change 02/06	Change 05/06
Iron	(t)	650.000	630.000	620.000	610.000	600.000	–7,69	–1,64
Nickel	(t)				5400	6400		18,52
Aluminium	(t)	380.100	389.100	397.500	394.200	367.400	–3,34	–6,80
Copper	(t)	1.200	600	1.400	7.900	8.700	625,00	10,13
Lead	(t)	6.200	1.800	0	0	0	–100,00	
Mercury	(t)	726	745	0	0	0	–100,00	
Tin	(t)	3	4	5	4	0	–100,00	–100,00
Zinc	(t)	64.900	15.100	0	0	0	–100,00	
Gold	(kg)	5.559	5.212	5.248	2.145	2.100	–62,22	–2,10
Silver	(kg)	12.800	2.200	3.600	5.200	5.300	–58,59	1,92
Baryte	(t)	41.000	50.000	52.000	42.000	35.000	–14,63	–16,67
Bentonite	(t)	100.000	225.000	250.000	162.500	110.000	10,00	–32,31
Diatomite	(t)	50.000	36.000	34.000	36.000	35.000	–30,00	–2,78
Feldspar	(t)	700.000	450.000	500.000	530.000	580.000	–17,14	9,43

Resources		2002	2003	2004	2005	2006	Change 02/06	Change 05/06
Fluorspar	(t)	134.000	125.000	130.000	130.059	132.000	–1,49	1,49
Gypsum	(t)	12.000.000	7.500.000	9.000.000	13.407.833	13.200.000	10,00	–1,55
Kaolin	(t)	515.000	580.000	600.000	728.188	750.000	45,63	3,00
Magnesite	(t)	185.000	150.000	130.000	120.000	200.000	8,11	66,67
Potash	(t)	407.000	510.000	600.000	500.000	437.000	7,37	–12,60
Salt	(t)	3.490.000	3.300.000	3.200.000	3.720.000	3.780.000	8,31	1,61
Sulfur	(t)	790.000	700.000	700.000	630.000	610.000	–22,78	–3,17
Talc	(t)	97.000	108.300	110.000	105.000	95.000	–2,06	–9,52
Steam-Coal	(t)	9.752.000	9.406.000	8.922.000	8.127.147	8.353.000	–14,35	2,78
Lignite	(t)	12.283.000	14.000.000	11.600.000	10.933.016	10.044.000	–18,23	–8,13
Nat. Gas	(Mm³)	551	549	540	535	530	–3,81	–0,93
Oil	(t)	324.000	310.000	300.000	300.000	290.000	–10,49	–3,33
Uranium	(t)	54	0	0	0	0	–100,00	

Remarks: Steam coal incl. anthracite

Table 140: Aggregates Production – Annual Statistics/Spain, 22 April 2008, quantities in million tonnes (UEPG, 2008)

Sand & Gravel (1)		Crushed rock (2)		Marine Aggregates		Recycled Aggregates (3)		Manufactured Aggregates (4)	
2004	2006	2004	2006	2005	2006	2005	2006	2005	2006
159,0	170,0	300,0	314,0	n. a.	0,0	1,3	1,5	0,0	0,0

(1) Sand and Gravel: sold production including crushed gravel
(2) Crushed rock: sold production (excluding crushed gravel)
(3) Recycled Aggregates: materials coming from construction and demolition waste used in aggregates market
(4) Manufactured aggregates include blast-furnace-slag, electric-arc-furnace-slag, incinerator bottom ash (IBA), pulverised fuel ash (PFA)

36.3 Normative Basics

36.3.1 Primary Legal Basics

The primary legal basics of mineral extraction activity is Mining Law No. 22 of 1973 as amended by Law No. 54 of 1980.

36.3.1.1 General Rules

Table 141: Structure of the Mining Law – Spain

Caption I Field (Scope) of application of the law and classification of the resources			Article 1: §§ 1–3
			Article 2: §§ 1–2
			Article 3: §§ 1–3
			Article 4: §§ 1–2
Caption II State action	Chapter I Accomplishment of studies, data summary and protection of the environment		Article 5: §§ 1–3
			Article 6: §§ 1–2
	Chapter II Areas reserved for the benefit of the state		Article 7
			Article 8: §§ 1–3
			Article 9: §§ 1–3
			Article 10
			Article 11: §§ 1–4
			Article 12: §§ 1–2
			Article 13: §§ 1–2
			Article 14: §§ 1–2
			Article 15: §§ 1–2
Caption III Regulation of the exploitation of resources of section A)			Article 16: §§ 1–2
			Article 17: §§ 1–3
			Article 18: §§ 1–2
			Article 19
			Article 20: §§ 1–2
			Article 21: §§ 1–3
			Article 22: §§ 1–3

Caption IV Regulation of the exploitation of resources of section B)	Chapter I Resources		Article 23. §§ 1–4
	Chapter II Authorizations of exploitation of resources of section B)	Section 1a Mineral and thermal water	Article 24. §§ 1–4
			Article 25: §§ 1–3
			Article 26: §§ 1–3
			Article 27: §§ 1–3
			Article 28: §§ 1–2
			Article 29
			Article 30
		Section 2a Deposits of non-natural origin	Article 31
			Article 32: §§ 1–2
			Article 31
		Section 3a Underground structures	Article 34: §§ 1–5
			Article 35: §§ 1–2
		Section 4a Compatibility of exploitation	Article 36: §§ 1–3
Caption V Regulation of the exploitation of section C)	Chapter I Free and registerable land		Article 37: §§ 1–2
			Article 38: §§ 1–2
			Article 39: §§ 1–3
	Chapter II Exploration permits		Article 40: §§ 1–2
			Article 41
			Article 42: §§ 1–2
	Chapter III Investigation permits		Article 43
			Article 44
			Article 45
			Article 46
			Article 47
			Article 48: §§ 1–4
			Article 49: §§ 1–2
			Article 50
			Article 51: §§ 1–3
			Article 52: §§ 1–3
			Article 53: §§ 1–3
			Article 54

			Article 55
			Article 56: §§ 1–4
			Article 57
			Article 58
			Article 59
	Chapter IV Exploitation	Section 1a General norms	Article 60
			Article 61
			Article 62: §§ 1–5
		Section 2a Direct concessions of exploitation	Article 63
			Article 64: §§ 1–2
			Article 65: §§ 1–3
			Article 66
		Section 3a Concessions of exploitation derived from investigation permit	Article 67
			Article 68: §§ 1–2
			Article 69: §§ 1–4
			Article 70: §§ 1–4
			Article 71: §§ 1–2
			Article 72
			Article 73: §§ 1–3
			Article 74: §§ 1–3
	Chapter V General conditions		Article 75: §§ 1–2
			Article 76: §§ 1–4
			Article 77
			Article 78: §§ 1–2
			Article 79
			Article 80: §§ 1–2
			Article 81
Caption VI Termination of files and cancellation of inscriptions			Article 82
Caption VII Expiration			Article 83: §§ 1–6
			Article 84: §§ 1–3
			Article 85: §§ 1–5
			Article 86: §§ 1–4
			Article 87: §§ 1–2
			Article 88

Caption VIII Conditions for usage of mining right titles		Article 89
		Article 90
		Article 91: §§ 1–2
		Article 92
		Article 93
Caption IX Transmission of mining rights		Article 94: §§ 1–4
		Article 95: §§ 1–5
		Article 96: §§ 1–2
		Article 97: §§ 1–3
		Article 98
		Article 99
		Article 100: §§ 1–2
		Article 101
Caption X Temporary occupation and enforced land expropriation		Article 102
		Article 103: §§ 1–2
		Article 104: §§ 1–4
		Article 105: §§ 1–4
		Article 106: §§ 1–4
		Article 107: §§ 1–2
Caption XI Mining boundaries		Article 108
		Article 109: §§ 1–3
		Article 110: §§ 1–4
		Article 111: §§ 1–2
Caption XII Establishment of gains (profits)		Article 112: §§ 1–3
		Article 113
Caption XIII Administrative competence and sanctions		Article 114: §§ 1–3
		Article 115: §§ 1–2
		Article 116: §§ 1–4
		Article 117: §§ 1–3
		Article 118
		Article 119
		Article 120
		Article 121: §§ 1–2
Final dispositions	Part 1–5	
Transitory dispositions	Part 1–10	
Additional disposition		

Mineral deposits are classified in the following sections (Article 3 ML):

- Section A: stone, sand and gravel etc.
- Section B: maritime or terrestrial mineral waters.
- Section C[1]: metallic ores and
- Section D: example coal, radioactive materials and other energy minerals.

Ownership of mineral rights

All mineral resources including territorial sea and continental shelf are public property, whose exploration and extraction rights the State represented by the Regional Government Administration can lease in the manner and conditions prescribed in this Law (Article 2 ML). The state can grant concessions (sections B, C, and D) or permits to extract (section A), this last to the land owner or anyone renting the land.[2]

Acquiring mineral rights

Holders of mineral rights can be natural or legal persons or foreign nationals (Article 78 ML). States or foreign companies can acquire mineral rights with the authorization of the responsible regional administration.

36.3.1.2 Issuing of Permits

Prospection and exploration

Regional administration grants prospection permits which confer their holders certain rights (Article 40 ML), for example conducting studies and surveys in selected areas. Prospection permits are granted for a period of one year, renewable for another year with an extension between 300 and 3000 hectares (Article 76 ML).

Regarding exploration permits concerning Section C minerals the following is required: Name and address of the applicant, designation of the land, municipality affected by the designation. Also a working and planning programme is required with indication of the techniques to use, and financing programme. Furthermore the applicant must present a general plan of investigation indicating the minerals of interest; he must specify the equipment and exploration budget and provide the results of an environmental impact assessment study.

---

1 Section C Mineral Resources: all mineral deposits and other geological resources not classified under the above sections or in section D.
2 Department of Environment (1994), lc.

The exploration permit is granted for three years. Its extension cannot surpass 300 ha. The permit can be extended until a maximum of 3 years.[3]

Exploitation

Section A minerals

Extraction of quarried minerals in terrains of private property relates to the ground owner (Article 16 ML). The approval to extract these resources, i. e. authorization of exploitation must be obtained from the competent regional authority (Consejerías Autonómicas). One must fulfill the following requirements:

- Indication about arrangement with the land owner when the mineral deposit is in private lands.
- Annual production, exploitation time and exploitation plan;
- Measures for environmental protection.
- The owner's permission to begin operation within six months of its notification grant (Article 18 ML).

Section C minerals

The right to use Section C) mineral resources is granted by the State by contract of concession (Article 60 ML). The competent authority to grant exploitation concession is the regional government. In order to get a concession the operator has to prove one or more resources susceptible for a rational use (Article 61 ML).

According to Article 62 ML the concession is granted for a period of 30 years renewable for periods equal to a maximum of 90 years. For an extension of the concession, the concessionaire must present (at least three years before the completion of the concession) a detailed report to the mining authority, in which he must demonstrate the continuity of the operated resource or the discovery of a new one, including a draft of operations for the next period.

Direct concessions of exploitation

In the following cases a direct exploitation permit can be obtained without preceding grant of an exploration permit (Article 63 ML): When a Section C resource already is proved, so it is considered sufficiently well known and its rational use is considered viable. The operator must submit a technical report

---

3   Espi, J. (2004), Country Report Spain, in: Department of Mining and Tunnelling, lc.

justifying the appropriateness of granting the concession. The following is required:

1) a designation of the land asked for;
2) a feasibility study describing the deposit including its reserves and resources;
3) a work program, list of facilities and machinery, and an approximate budget regarding the extraction of the deposit;
4) an environmental impact assessment study;
5) Economic study regarding financing and its viability guarantee.

Concession exploitation derived from exploration permissions

As soon as the exploration proves sufficiently the existence of a Section C resource, the holder may ask for a permit to operate all or part of the explored area (Article 67 ML).

In accordance with Article 68, the concession is granted by competent regional authorities at provincial level. Beside other items, the application must include the working programme and estimated budget (Article 67). If the documents submitted meet the regulatory requirements and the project has been deemed appropriate for rational use of the resource indented by exploration, the Provincial Delegation of the regional administration verifies the application area.

The competent authority (at provincial level) will grant or refuse exploitation concession and may impose special conditions as it deems appropriate, including appropriate environmental protection (Article 69 ML). The holder must begin work within one year from the date of grant and has to submit the working plan and facilities to perform the mining activities in the first year to the Provincial Delegation of regional administration within six months from the date of grant (Article 70 ML).

To increase use of resources, the regional administration can encourage the establishment of various mining concessions as a grouping of interests of holders of exploration rights in certain areas, so as to allow joint use of the required services. The administration may provide tax benefits and other incentives (Article 108 ML).

36.3.1.3 Authorities

The main responsible authority for mining is the General Directorate of Regional Administration Authority, the regional administration in areas with competence in exploration and mining exploitation (Article 114 ML). Responsibility for controlling the mining activities has been delegated from state to regional level, i. e. to the regional Departments of Industry and Environment (which are responsible for issuing of permits.[4]

36.3.1.4 Fees and Taxes

Table 142: Taxes according Article 66 of Decree No. 88 of 1990

Articles	Assignments	Taxes
Article 8	Prospection and exploration contract	
	Metallic minerals	€ 150
	Non metallic minerals	€ 75
File d) n 1 of article 8	Exploration period	€ 50
Article 11	Transmission exploration contractual position.	
	Metallic minerals	€ 125
	Non metallic minerals	€ 50
Article 20	Experimental concession contract	€ 200
Article 21	Concesión contract.	€ 325
File c) article 21	Registering new substance to the concession contract.	€ 150
File d) article 21	Concession contract extension	€ 50
Article 22	Concession transmission	€ 250
Article 23	Concession landmark area	€ 150
Article 24	Alteration of concession area	€ 150
Article 25	Voluntary integration of concessions for each concession.	€ 150
N 3 article 27	Work Plan alteration	€ 100

Table 143: Taxes according to Article 56 of Decree No. 89 of 1990 (construction minerals):

Articles	Designation	Taxes
File a) n 2 article 20	Work Plan alteration	€ 75

---

4   Department of Environment (1994), lc.

36.3.2 Additional Legal Basics

Table 144: List of Acts significant for raw materials – Spain (Espi, 2004)

Name of Law	No of Law and Year of Issuing
Promotion of the Mining Law	Law No.6 of 1977
Modification Law 22/73	Law No.54 of 1980
Mining Status Royal Decree	Royal Decree No. 3255 of 1983
General Regulation of Mining Basic Safety Rules	Royal Decree No.863 of 1985
Restoration of Areas Affected by Mining Activities	Royal Decree No.2994 of 1985
Modification of Royal Decree 1302/1986 of Environmental Impact Assessment	Law No 6/2001 of 1986
General Regulation Decree for the Mining Law	Decree No.2857 of 1987
Conservation of Natural Areas and Wildlife	Law No. 4 of 1989
Minimum safety rules to protect mine workers	Royal Decree 1389 of 1997
Amendment of 34/1998 Hydrocarbons Law	Law 12 of 2007

37 Sweden

37.1 General Facts

The national territory covers 449 964 km² and the inhabitants are 9,2 million. The population density is 21/km². In 2009 the GDP per capita amounted to $ 43.986.

Constitutional Structure

The form of government is a constitutional monarchy. Sweden is divided into the national level, district level (21 districts), and municipal level (289 municipalities). The national government is in each district represented by a County Administrative Board (Länsstyrelsen) and Country Governor (Länshövding).

37.2 Production of Raw Materials

Table 145: Production Date – Sweden (World Mining Data, Weber and Zsak, 2008)

Resources		2002	2003	2004	2005	2006	Change 02/06	Change 05/06
Iron	(t)	12.262.000	13.300.000	14.700.000	14.720.000	14.800.000	20,7	0,54
Aluminium	(t)	101.100	100.700	100.600	102.107	101.700	0,59	–0,40
Copper	(t)	71.991	83.100	82.400	87.068	86.746	20,50	–0,37
Lead	(t)	42.954	50.962	54.347	60.445	55.644	29,54	–7,94
Zinc	(t)	148.620	189.700	197.000	215.700	210.029	41,32	–2,63
Gold	(kg)	5.757	5.900	6.564	5.163	6.848	18,95	32,64
Silver	(kg)	320.823	340.700	319.600	309.933	292.255	–8,90	–5,70
Feldspar	(t)	37.000	35.000	42.000	38.000	42.000	13,51	10,53
Talc	(t)	20.000	22.000	14.000	12.000	20.000	0,00	66,67

Table 146: Aggregates Production – Annual Statistics/Sweden, 22 April 2008, quantities in million tonnes (UEPG, 2008)

Sand & Gravel (1)		Crushed rock (2)		Marine Aggregates		Recycled Aggregates (3)		Manufactured Aggregates (4)	
2005	2006	2005	2006	2005	2006	2005	2006	2005	2006
23,0	23,0	49,0	62,0	n.a.	0,0	7,9	1,8	0,2	0,2

(1) Sand and Gravel: sold production including crushed gravel
(2) Crushed rock: sold production (excluding crushed gravel)
(3) Recycled Aggregates: materials coming from construction and demolition waste used in aggregates market
(4) Manufactured aggregates include blast-furnace-slag, electric-arc-furnace-slag, incinerator bottom ash (IBA), pulverised fuel ash (PFA)

37.3 Normative Basics

37.3.1 Primary Legal Basics

The primary legal basics of mineral extraction activity is the Mining Law No. 45 of 1991 as amended by Law No. 943 of 2005 (concession minerals) and the Minerals Ordinance No. 285 of 1992, last amended by Law No. 162 of 2005. Aggregates are not covered by the Minerals Act: Non concession minerals are mainly regulated by the Environmental Code No. 808 of 1998.

37.3.1.1 General Rules

Table 147: Structure of the Mining Law – Sweden[1]

Chapter 1: Introductory Regulations
Application of the law Articles 1–3
The right to exploration and exploitation Articles 4–6
Provisions in other legislations Article 7
Chapter 2: Exploration permit
Permit area Article 1
Prerequisites for a permit Article 2
Precedence in the case of competition Article 3 – Article 4
Period of validity for an exploration permit Articles 5–8
Impediments concerning the granting of an exploration permit Articles 9, 9a, 9b

---

1 Notification: Each chapter of the Mining Law starts with 'Article 1'.

Conditions in certain exploration permits Article 10

Chapter 3: Exploration Work
Introductory provisions Article 1

The right to explore without a permit Article 2

How the exploration work may be carried out Articles 3–5

Impediments to exploration work Articles 6–8

Chapter 4: Exploitation Concession
Concession area. Article 1

The prerequisites for concession Article 2

Precedence in the case of competition etc. Articles 3–6

Period of validity for an exploitation concession Articles 7–11

Chapter 5: Exploitation etc.
Introductory provisions Article 1

The right to carry out exploitation work without concession. Article 2

How exploration work may be conducted Article 3

How exploitation work etc. may be conducted Articles 4–7

Transport roads and fencing etc. Articles 8–9

Impediments to exploration work and exploitation etc. Articles 10–11

Chapter 6: Transfer, Relinquishing, Revocation and Amendment of Exploration Permits and Exploitation Concessions
Transfer Article 1

Relinquishing Article 2

Revocation Article 3

Amendments of conditions Article 4

Chapter 7: Compensation
Compensation for plaintiff Articles 1–4

Compensation in connection with revocation Article 5

Chapter 8: Authorities Responsible for Examination etc.
Exploration permit and exploitation concession Articles 1–6, 6a

Examination of certain disputes Articles 7–11

Costs in certain disputes Article 12

Chapter 9: Proceeding for Designation of Land
General provisions Articles 1–3

The authority for the designation proceeding etc. Articles 4–10

Application for designation of land Articles 11–12
Meeting with the interested parties etc. Articles 13–18
Voting Article 19
The costs of the proceeding Articles 20–21
The content of the decision about designation of land Articles 22–23
Time for announcement of decision Article 24
Chart, Map, etc. Article 25
Access to land Article 26

Chapter 10: Payment of Compensation etc.
Articles 1–7

Chapter 11: Excluded since 1st of July 1993

Chapter 12: Joint Administration of Concession Rights
Introductory provision Article 1
Manager for the operations Articles 2–3
Meeting with the share-holders Articles 4–5
Exploration and exploitation Articles 6–8
Forfeiture of a share in the concession Article 9

Chapter 13: Effect of Termination of an Exploitation Concession
Articles 1–8

Chapter 14: Fees and other Special Obligations
Fees Articles 1–2
Furnishing reports on exploration works carried out Article 3
Obligation to produce a chart and to keep records etc. Articles 4–5

Chapter 15: Supervision, Assistance, Penalties and Responsibility
Articles 1–6

Chapter 16: Appeal against Decisions
Articles 1–6

Chapter 17: Special Provisions
Articles 1–3

Chapter 1 – Introductory Regulations

The Minerals Act covers a number of minerals called concession minerals, and these belong to the state, regardless of who owns the land where they occur. The concession minerals are divided into three groups (Article 1 ML):

1) antimony, arsenic, beryllium, lead, cesium, gold, iridium, iron (in the bedrock), cobalt, copper, chromium, mercury, lanthanum and lanthanides, lithium, manganese, molybdenum, nickel, niobium, osmium, palladium, platinum, rhodium, rubidium, ruthenium, silver, scandium, strontium, tantalum, tin, titanium, thorium, uranium, vanadium, bismuth, tungsten, yttrium, zinc and zirconium,

2) alum shale, andalusite, apatite, brucite, fluorspar, graphite, kyanite, refractory clay or clinkering clay, magnesite, pyrrhotite, nefelinsyenite, sillimanite, coal, rock salt or other salt present in a similar manner, pyrite, baryte and wollastonite,

3) oil, gaseous hydrocarbons and diamonds.

This Act is not applicable to areas within public waters of the sea.

All other mineral substances belong to the landowner.

Ownership of minerals

The state has the responsibility about issuing mineral rights (concession minerals). The right to explore for and exploit non-concession minerals is regulated by contract with the landowner.

37.3.1.2 Issuing of Permits

Exploration
Chapter 2 – Exploration permit

An exploration permit applies to a specific area. The permit holder must explore in an appropriate manner and its size must also be suitable for the intended purpose (Article 1 ML). An exploration permit is granted, if 1) there is reason to assume that exploration in the area can lead to the discovery of a concession mineral and 2) the applicant does not obviously lack the possibility or intention of conducting appropriate exploration (Article 2 ML).

A permit for exploration with regard to diamonds can only be granted to a person who can prove that he is suitable to carry out such exploration work.

If there are several applications for exploration permits within the same area, the applicant who first submitted an application has precedence. If applications are received on the same day, the applicants have equal rights in respect of the area which they have in common (Article 3 ML).

If a person holds an exploration permit or an exploitation concession, no other person can be granted an exploration permit for the same mineral within this area. If special reasons apply, some other person can be granted an exploration permit within the area for minerals not covered by the permit or the concession (Article 4 ML).

An exploration permit is valid for 3 years from the date of the decision (Article 5 ML).

The period of validity for the exploration permit can be extended on application by the permit-holder for a total of not more than 3 years, if appropriate exploration work has been carried out within the area. The same also applies if the permit-holder has acceptable reasons for not operator exploration work and, furthermore, that it appears probable that the area will be explored in the period covered by the application (Article 6 ML).

If the period of validity for the exploration permit has been extended under Article 6 ML, the permit can be extended on application by the permit-holder by a further total of not more than 4 years, if special reasons apply (Article 7 ML). Subsequently, the period of validity can be extended by a further period of not more than 5 years if extraordinary reasons apply, for instance if the permit-holder shows that considerable work has been undertaken in the area and that further exploration will probably result in the granting of an exploitation concession.

Applications in respect of extension of the period of validity can be received during the period of validity if they are to be considered. If the permit-holder has applied for extension within the stated period or if he has applied for an exploitation concession within the same period, the permit is valid until the application has been finally decided on (Article 8 ML).

Chapter 3 – Exploration Work

According to Article 1 ML a person holding an exploration permit can carry out exploration work within the permit area (in accordance with Articles 3–8 ML).

Exploration work can only be carried out to show that a mineral covered by the permit is present within the area and to obtain more knowledge about the size, nature and extractability of the deposit. The permit-holder can utilize land in order to erect buildings which are necessary for the exploration work. Where necessary, he can also utilize land in order to build a road within the area or use existing roads to and within the area. The permit-holder can also utilize land in order to build an essential road leading to the area, on receipt of permit from the Mining Inspector. The work must be so carried out as to cause the least possible damage and encroachment (Article 3 ML).

The permit-holder can only use concession minerals which are extracted during operations and which are covered by the permit to study their character and suitability for technical processing. (Article 4 ML).

At least two weeks before exploration work commences, the permit-holder must give notification of commencement of the work to the Mining Inspector as well as the owner of the land on which the work is to be carried out and to the holders of usufructuary rights or easements. If it can be assumed that the operations will presumably affect other special rights, the holder of such rights must be notified (Article 5 ML).

Exploitation
Chapter 4 – Exploitation Concession

A concession can be valid for a definite area, which is decided on the basis of what is appropriate with regard to the extent of the deposit, the purpose of the concession and other circumstances (Article 1 ML). A concession may be granted if 1) a deposit has been found which can probably be utilized on an economic basis, 2) the location and nature of the deposit does not make it inappropriate that the applicant is granted the concession requested (Article 2 ML).

If there are several applications for a concession within the same area and more than one person can be taken into consideration under Article 2 ML, the person who holds an exploration permit within the area for a mineral which is covered by his application for a concession will have precedence (Article 3 ML). If none of the applicants holds an exploration permit, the applicant who has carried out appropriate exploration work within the area will have precedence. Otherwise the applicant who first submitted an application has precedence. If the applications are received on the same day, the applicants shall have equal right to share in the concession.

If a person holds an exploration permit nobody else can be granted a concession for the same mineral within this area. If special reasons apply, another person can be granted a concession within the area for minerals which are not covered by the permit or the concession (Article 4 ML). A concession has to comply with the conditions required to protect the public interest, or individual rights, or which are required for appropriate exploration and exploitation of the natural resources (Article 5 ML).

Period of validity for an exploitation concession: An exploitation concession is valid for 25 years. A shorter period can be decided on request of the applicant (Article 7 ML). The concession period can be extended by ten years at a time without special application if regular exploitation is continuing when the period of validity expires. If the concession-holder so requests, a shorter period may be decided (Article 8 ML).

If regular exploitation work is not in progress when the concession expires, the period can be extended for ten years, following an application from the concession-holder, provided that there is in progress 1) preparatory work or construction work for an exploitation within the concession area 2) exploration work on a large scale, within the concession area, or 3) mineral-technical, metallurgical or other development work of greater extent with purpose to make it possible to exploit the deposit (Article 9 ML).

Chapter 5 – Exploitation, further requirements

A person holding a concession can within the concession area, on the one hand carry on exploration work in accordance with Article 3, and on the other hand execute exploitation work and activities connected therewith in accordance with Articles 4–7 ML. However, the concession-holder can only make use of land which has been assigned for this purpose, for exploitation work and activities connected therewith which are carried out in the open (Article 1 ML). Land outside the concession area can be used for other purposes than exploration work or exploitation work in accordance with the decision regarding the designation of the land. Matters regarding land designation must be considered in proceedings relating to land designation in accordance with the provisions contained in Chapter 9 of the Mining Law.

An exploration permit or an exploitation concession can be transferred following the consent of the authority responsible for examination. (Chapter 6 Article 1 ML).

Restoration

This is repulated by the Environmental Code.

Any new operation, will in its application for an extraction permit, also outline the plans for restoration and aftercare. Such work should as far as possible be an integral part of the daily operations, and not be postponed to the final phase of the project. New operations must give a financial guarantee to secure future restoration. The same stipulations apply to non-concession minerals.[2]

---

2   Nielsen (2004) lc.

Quarries

Permission can be granted from the County Administrative Board according to new Rules in Chapter 9 of the Environmental Code. According to the Environmental Code permit must be obtained from the county administrative board for the quarrying of rock, stone, gravel, sand, clay, soil or peat. In connection with the consideration of applications for quarrying permits, the demand for the material to be extracted shall be balanced against the damage that the quarrying is likely to cause to wild flora and fauna and the environment in general. The need of cultivable agricultural land shall be taken into account in connection with the consideration of applications for permits for the quarrying of topsoil.

Quarrying permits may only be granted if a security is furnished for fulfilment of the conditions attached to the permit. In special circumstances, the county administrative board may waive the requirement to furnish a security. Further provisions relating to securities are contained in the Environmental Code. The owners or other possessors of the land have to accept any measures that are carried out in accordance with the conditions attached to the quarrying permit.

The Government or the authority appointed by the Government may issue rules to the effect that operators of quarries for which a permit is required pursuant to this Code or to rules issued in pursuance thereof or operators of stone-crushing plants shall supply any information about the operations that may be required as guidance for nature conservation planning. The Government may require holders of quarrying permits to notify the county administrative board of the name of the quarry operator. Where the operator of a quarry for which a permit has been granted is not known, the holder of the permit shall be deemed to be the operator for the purposes of application of this section and of any rules issued in pursuance thereof.

If an activity or measure for which a permit or notification is not required pursuant to other provisions of this Code is liable to have a significant impact on the natural environment, notice of consultation shall be made to the supervisory authority.

The Government or the authority appointed by the Government may issue rules to the effect that notice of consultation shall always be given, at national or local level, in the case of activities or measures that may damage the natural environment. The Government or the authority appointed by the Government may also issue rules specifying the information to be supplied in such a notice. Activities or measures which must be notified for consultation may not commence earlier than 6 weeks after submission of the notice unless the supervisory authority decides otherwise.

37.3.1.3 Authorities

Chapter 8 – Authorities Responsible for Examination[3]

The main responsible authority for mining is the Minister of Industry (Mining Inspectorate).

Matters relating to the granting of an exploration permit or an exploitation concession will be considered by the competent authority, i. e. mining inspector (Ministry of Industry). In matters relating to the granting of an exploitation concession, the Mining Inspector must consult the County Administrative Board in the county or counties where the concession is situated as regards application of Chapters 3, 4 and 6 of the Environmental Code (Article 1 ML).

Cases relating to the granting of an exploitation concession must be referred for consideration by the Government if 1) the mining inspector considers the matter of concession to be of special importance from point of view of the public interest or 2) the mining inspector, in applying Chapter 3 or 4 of the Environmental Code, finds reason to reject what the county administrative board has proposed (Article 2 ML).

Chapter 9 – Proceedings for Designation of Land

A proceeding for designation of land must be held at the request of the concession holder. In the proceeding, the land within the concession area which the concession holder may utilize for the exploitation of a mineral deposit must be determined. A determination must also be made of the land or space, within or outside the concession area, which the concession holder may use for activities connected with the exploitation of the deposit. The nature of such activities must be stated.

Chapter 15 – Supervision, Assistance, Penalties and Responsibility

The mining inspector must exercise supervision of compliance with this Law (Article 1 ML). A person who holds an exploration permit or an exploitation concession and others carrying on activities covered by this Act has at the request of the mining inspector (Article 2 ML) 1) to provide the mining inspector with such information, and 2) give the Mining Inspector access to installations or places where activities covered by this Act are carried on.

---

3   According to the Minerals Act.

The mining inspector can issue regulations to secure compliance with this Act or with conditions or provisions which are issued under the Act. If work is carried on in such a manner that evident danger to public or private interests arises, the Mining Inspector may prohibit the work. Such prohibition takes immediate effect and may be implemented (Article 3 ML).

37.3.1.4 Fees and Taxes

Chapter 14 – Fees

An applicant must pay an application fee in accordance with the Government's stipulations (Article 1 ML). A holder of an exploration permit must pay a fee to the State. The fee must be decided with regard to the extent of the exploration area and the nature of the minerals which are covered by the permit in accordance with what the Government further prescribes (Article 2).

37.3.2 Additional Legal Basics[4]

Table 148: List of Acts significant for Raw Materials – Sweden

Name of Law	No. of Law and Year of Issuing
Environmental Code	Law No. 808 of 1998
Planning and Building Act	Law No. 900 of 2010
Ancient Monuments and Finds Act	Law No. 850 of 1988
Continental Shelf Act	Law No. 314 of 1966
Ground Water Exploration Act	Law No. 424 of 1975
The Administrative Procedure Act	Law No. 223 of 1986
Land Code)	Law No. 994 of 1970
Real Property Formation Act	Law No. 988 of 1970
Real Property Register Act	Law No. 224 of 2000
Swedish Competition Act	Law No. 729 of 1982
The Swedish Work Environment Act	Law No. 1160 of 1977
Forest Act	2004
Ordinance concerning Environmentally Hazardous Activities and The Protection of Public Health	Law No. 899 of 1998

---

4   Nielsen (2004), lc.

Environmental Code

The purpose of this Code is to promote sustainable development which will assure a healthy and sound environment for present and future generations. Such development will be based on recognition of the fact that nature is worthy of protection and that the right to modify and exploit nature carries with it a responsibility for wise management of natural resources (Article 1).

The Environmental Code covers not only environmental aspects but also issues related to water supply, nature conservation and recycling. The Environmental Code must be applied in such a way as to ensure that:

1. human health and the environment are protected against damage and detriment, whether caused by pollutants or other impacts;
2. valuable natural and cultural environments are protected and preserved;
3. biological diversity is preserved;
4. the use of land, water and the physical environment in general is such as to secure a long term good management in ecological, social, cultural and economic terms; and
5. reuse and recycling, as well as other management of materials, raw materials and energy are encouraged with a view to establishing and maintaining natural cycles.

In matters concerning granting a mining concession (i. e. concession minerals), Chapters 3 and 4 of the Environmental Code must be applied. If the granting of a concession involves an operation which is to be subject to subsequent review in accordance with the Environmental Code, only Chapters 3 and 4 of the Environmental Code are to be applied in examining the question of the concession. In granting a concession, an environmental impact assessment has to be appended to the application.[5]

The Ground Water Exploration Act No. 424 of 5 June 1975 does not influence the mineral sector directly, but the act stipulates that all water exploration and well drilling contractors and others shall submit a report to the Swedish Geological Survey about the results, including information about the soils and rock encountered during the work. The Ground Water Exploration Act is administered by the Ministry of Industry, Employment and Communications.

---

[5] Chapter 6, Article 3, Article 7, Article 8 first paragraph, and Articles 9–12 of the Environmental Code are to be applied as assessment, plans and planning documentation.

38 Switzerland

38.1 General Facts

The national territory covers 41,284 km² and the inhabitants are 7,5 million. The population density is 182/km². The GDP per capita for 2009 is estimated at $ 69.838.

Constitutional structure

The form of government is a federal state with 26 cantons.

38.2 Production of Raw Materials

Table 149: Production Data – Switzerland (World Mining Data, Weber and Zsak, 2008)

Resources		2002	2003	2004	2005	2006	Change 02/06	Change 05/06
Aluminium	(t)	40.000	43.900	44.538	44.800	12.000	–70,00	–73,21
Gypsum	(t)	300.000	300.000	300.000	300.000	300.000	0,00	0,00
Salt	(t)	434.000	562.000	569.000	566.000	560.000	29,03	–1,06

Table 150: Aggregates Production – Annual Statistics/Switzerland, 22 April 2008, quantities in million tonnes (UEPG, 2008)

Sand & Gravel (1)		Crushed rock (2)		Marine Aggregates		Recycled Aggregates (3)		Manufactured Aggregates (4)	
2005	2006	2005	2006	2005	2006	2005	2006	2005	2006
46,5	50,0	5,3	5,7	n. a.	0,0	5,3	5,7	n. a.	n. a.

(1) Sand and Gravel: sold production including crushed gravel
(2) Crushed rock: sold production (excluding crushed gravel)
(3) Recycled Aggregates: materials coming from construction and demolition waste used in aggregates market
(4) Manufactured aggregates include blast-furnace-slag, electric-arc-furnace-slag, incinerator bottom ash (IBA), pulverised fuel ash (PFA)

38.3 Normative Basics

38.3.1 Primary Legal Basics

In Switzerland the cantons are authorized to pass legal regulations with regard to the Mining Law in an autonomous way. Nowadays the National Constitution does not contain a special assignment of responsibilities for the matters of the Mining Law. According to Article 3 of the National Constitution the federal state can only take actions in areas for which it was given the necessary responsibilities by the National Constitution. The National Civil Law more or less only mentions matters of Mining Law in connection with proprietary, namely in Article 655 and Article 943 National Civil Law.

Mining Law basics differ a lot from canton to canton.[1] Because of the *variety of mining laws* of the cantons the sections below deal *only* with the Mining Law of the canton *Bern* ("Gesetz über die Gewinnung mineralischer Rohstoffe", adopted in 1979).

Construction minerals are regulated by the Land Use Planning Law and Environmental Law.

38.3.1.1 General Rules

Ownership of mineral rights

Mineral resources usually are divided into 2 categories. Free for mining (or "regale") minerals: The ownership of land does not apply to the free for mining minerals that are listed in the mining law. Free for mining minerals include all mineral raw materials associated with iron; all types of coal and oil shale; rock salt and all other salts. As these minerals are not owned by anybody, the rights to explore and/or to extract these minerals remain separate from the ownership of land. To acquire these rights, a mining concession must be obtained (Article 1 ML).

Minerals owned by the landowner: The minerals, which are not listed in the mining law, belong to the landowner.

---

[1] Kündig et al, (1997), Die mineralischen Rohstoffe der Schweiz, Zürich: A summary (1990) shows the following: no cantonal legal regulations can be found in the cantons Basel Stadt, Appenzell Innerrhoden and Graubünden. The cantons Uri (1984), Obwalden (1968), Nidwalden (1990), Glarus (1988), Basel Landschaft (1984), Aargau (1980) and Thurgau (1990) have National Constitution regulations regarding the Mining Law. Some cantons have adopted Mining Law regulations to the cantonal introduction law ("Einführungsgesetz") as part of the Civil Code of Laws, namely Zürich (1911), Nidwalden (1990), Zug (1942), Schaffhausen (1911) and Appenzell Ausserrhoden (1969).

38.3.1.2 Issuing of Permits

Exploration

The land owner is entitled to explore on his property after a prior report to the department of building, transport and energy, as long as an exploration permit has not been given to a third party. The department of building, transport and energy can order protection measures or forbid the exploration under Article 27 ML. In all other cases the exploration of mineral resources requires a permit.

The exploration permit ("Schurfbewilligung") authorizes the approval holder to explore a certain area (Article 23 ML). Before starting the exploration a detailed exploration program must be approved by the department of building, transport and energy. The affected municipalities must be heard before the approval of the program (Article 28 ML).

The holder of the exploration permit must submit annually a detailed report with precise information about the carried out and the still planned work as well as about the results to the responsible department. A final report should be submitted after the expiration of the preliminary exploration permit (Article 30 ML).

The exploration permit is given for the period of time of at most 3 years. It can be extended in each case by a year (Article 25 ML).

The exploration permit holder is entitled to use the mineral resources exploited in such a way as it is necessary for the clarification of the deposits exploitation worthiness. On instruction of the Department of Building, Transport and Energy, the prospector must submit reports, work reports, plans, maps, profiles and rock samples. If the holder of an exploration permit can prove a workable deposit, he has the right to obtain a mining permit ("Ausbeutungsbewilligung") according to Article 38 ML.

Exploitation

By granting a mining permit (Articles 39 to 52 ML), the concessionaire is given the right to exploit mineral resources, which are subject to state sovereignty, on certain properties.

The applicant must submit an application to the Department of Building, Transport and Energy. Besides information about the applicant and the property in question, it must specifically include proof of exploitation worthiness, proof about the taking out of sufficient public liability insurance and proof regarding financing. Additionally the application must include a working programme as well as a location map (official excerpt from the land survey office).

The application is publicly displayed at the local authority office ("Regierungsstatthalteramt") and published in the gazette. The local authority office must inform the affected land owners additionally by a registered letter. Within 30 days an objection can be raised against the applied permit because of a violation of private or public interests (Article 41 ML).

Afterwards the Department of Building, Transport and Energy examines the application and the objections in agreement with the Department of Justice, local authorities and church offices. After the termination of examinations, the Department of Building, Transport and Energy transfers the files with the application to legislature that decides on the application and the objections of a public nature.

The permit is granted for a maximum period of 50 years (Article 46 Mining Law). A mining right given for at least 30 years can be recorded in the land register as an independent and permanent right. Legislature can renew an expired permit and simultaneously place new conditions. The renewal must be granted, as far as no public interests are in the way and no interests of a third party are severely violated.

The land owner is entitled to a full compensation of damage of property, of decrease in profits as well as of all other disadvantages that arose for him due to the exploitation. For the land owner's entitlement to damage compensation and for the costs of restoring the previous condition, the one given approval or a permit of extraction must provide for the payment an appropriate compensation, the amount is fixed by the Department of Building, Transport and Energy.

The land owner can ask the concessionaire to buy the property, if the landowner has been withdrawn the use and cultivation of the land for more than three years or if the land has become permanently unusable for the previous cultivation. In case of dispute, the judge of dispossession makes the decision about the obligation of takeover and the price.

Article 11 ML regulates the restoration. After termination of exploration and exploitation the previous condition of the property must be restored in as accurate a way as possible. The measures to be taken are defined by the Department of Building, Transport and Energy.

38.3.1.3 Authorities

The monitoring for minerals covered by the Mining Law is the responsibility of the Department of Building, Transport and Energy.

38.3.1.4 Fees and Taxes

Explorers and concessionaires as well as their legal successors must pay the state the following fees: administrative fees, permit fees, production fees. For solid mineral resources the annual production fee is 10 % of the entire exploitation's current market value. The cantonal parliament passes relevant basic approaches and required regulations about the amount and the payments of all the other fees in a decree (Articles 53–57 ML).

38.3.2 Additional Legal Basics

Table 151: List of acts significant for raw materials – Switzerland

Name of the Law	No. of Laws and Year of Issuing
Federal Law on the Protection of Environment	1983
Federal Land Use Planning Law	1979
Federal Law on the Protection of Water	1991
Federal Law on the Conservation of Nature	1966

Table 152: Legal basics – Swiss construction minerals industry (Grob, 2005)

	Legal and administrative Framework	Exploration companies	Target	Remarks
Land use planning stage	Federals laws: (Land Use Planning Law) Supply of raw materials are of public interest, assignment for supply, obligation for a raw material strategy, comprehensive consideration of interests	Swiss (minerals) Associations campaign for proper application of the concept and prevent large-scale exclusion zones. Cantonal and regional organizations participate in the concepts,	best possible mining site	A comprehensive consideration of interests enables to create a balance for the three pillars of the sustainability principle. The emphasis may be varying. For rare raw materials the protection of the deposits may be most important (economic and social priority), for sand and gravel, the priority is rather to find the best extraction

38.3 Normative Basics

	Legal and administrative Framework	Exploration companies	Target	Remarks
Land use planning stage	Cantonal laws: raw material strategy as part of land use planing, partly cantonal concerned plans Municipal planning: Mining zone as part of local planning, partly special usage plans	enterprises partipate on site selection Enterprises develop projects based on environmental impact assessment.	best possible mining site	site (economic and environmental priority) Raw materials planning process can be linked with other planning objectives (infrastructure, nature habitats) and thus create a balance in terms of sustainability.
Permission stage	Federal laws: Environment law (and further) regulations of special fields, guidelines, communications Cantonal laws: enforcement law and others Local authorities partial enforcement	FSKB (Fachverband der Schweizerischen Kies- und Betonindustrie) Inspectorate as a self-regulatory mechanism, on the basis of laws and regulations, branch guidelines and standards. Used by cantonal and municipal authorities as controlling body. Integrable in ISO certification for quality, environment and operational safety.	best practice	Strong focus on ecological aspects. The sustainability concept can be taken into account if the use of "technically feasible and economically viable" is applied reasonably.

39 Turkey

39.1 General Facts

The national territory covers 783 562 km² and the inhabitants are 74.616.000 (2009 estimation). The population density is 95,5/km². The GDP per capita for 2009 is estimated at $ 8.723.

Constitutional Structure

The form of government is a parliamentary republic. It is subdivided into 81 provinces. These are organized into 7 regions for census purposes. Each province is divided into districts, in total 923 districts.

39.2 Production of Raw Materials

Table 153: Production Data – Turkey (World Mining Data, Weber and Zsak, 2008)

Resources		2002	2003	2004	2005	2006	Change 02/06	Change 05/06
Iron	(t)	2.600.000	2.297.477	2.583.879	2.606.256	2.800.000	7,69	7,43
Chromium	(t)	160.000	118.349	183.388	289.118	310.000	93,75	7,22
Nickel	(t)	0	500	0	700	2.600	–	271,43
Aluminium	(t)	62.500	62.900	62.400	59.000	60.000	–4,00	1,69
Antimony	(t)	20	20	20	0	0	–100,00	–
Bauxite	(t)	287.400	207.654	208.527	203.194	771.200	168,34	279,54
Copper	(t)	60.000	59.000	49.300	54.100	46.400	–22,67	–14,23
Lead	(t)	17.300	17.400	18.700	19.000	18.000	4,05	–5,26
Zinc	(t)	43.000	40.000	39.000	56.000	62.000	62000	10,71
Gold	(kg)	5.757	5.900	6.564	5.163	6.737	17,02	30,49
Silver	(kg)	124.000	158.000	122.000	219.000	220.000	220000	0,46
Baryte	(t)	160.000	119.648	134.504	70.925	180.000	12,50	153,79
Bentonite	(t)	680.000	831.146	643.153	582.735	570.000	–16,18	–2,19
Boron	(t)	1.300.000	1.280.113	1.669.779	2.017.402	2.500.000	92,31	23,92

Resources		2002	2003	2004	2005	2006	Change 02/06	Change 05/06
Feldspar	(t)	700.000	1.199.328	1.336.768	1.571.748	1.550.200	121,46	−1,37
Fluorspar	(t)	1.500	718	880	900	800	−46,67	−11,11
Graphite	(t)	1.500	942	900	850	1.200	−20,00	41,18
Gypsum	(t)	270.000	300.000	250.099	24.421	30.000	−88,89	22,85
Kaolin	(t)	300.000	370.455	536.008	615.271	580.000	93,33	−5,73
Magnesite	(t)	980.000	1.130.000	1.300.000	2.371.206	3.200.000	226,53	34,95
Perlite	(t)	200.000	137.313	133.829	156.935	145.000	−27,50	−7,61
Salt	(t)	1.600.000	1.400.000	2.157.718	1.726.233	1.600.000	0,00	−7,31
Sulfur	(t)	80.000	82.000	85.000	90.000	93.000	16,25	3,33
Talc	(t)	4.700	4.900	5.000	5.300	7.000	48,94	32,08
Steam-Coal	(t)	1.570.000	1.320.000	1.220.000	1.520.000	2.300.000	46,50	51,32
Coking-Coal	(t)	670.000	690.000	660.000	650.000	930.000	38,81	43,08
Lignite	(t)	51.670.000	43.749.420	43.754.159	60.857.574	61.640.000	19,30	1,29
Nat. Gas	(Mm^3)	268	276	344	484	490	82,84	1,24
Oil	(t)	2.400.000	2.375.082	2.275.529	2.280.764	1.800.000	−25,00	−21,08

Table 154: Aggregates Production – Annual Statistics/Turkey, 22 April 2008, quantities in million tonnes (UEPG, 2008)

Sand & Gravel (1)		Crushed rock (2)		Marine Aggregates		Recycled Aggregates (3)		Manufactured Aggregates (4)	
2004	2006	2004	2006	2005	2006	2005	2006	2005	2006
n.a.	240,0	n.a.	260,0	n.a.	0,0	n.a	0,0	n.a.	0,0

(1) Sand and Gravel: sold production including crushed gravel
(2) Crushed rock: sold production (excluding crushed gravel)
(3) Recycled Aggregates: materials coming from construction and demolition waste used in aggregates market
(4) Manufactured aggregates include blast-furnace-slag, electric-arc-furnace-slag, incinerator bottom ash (IBA), pulverised fuel ash (PFA)

39.3 Normative Basics

39.3.1 Primary Legal Basics

The primary legal basics of mineral extraction activity is the Mining Law No. 3213 of 1985, as amended by Law No. 5177 of 2004.

39.3.1.1 General Rules

Table 155: Structure of the Mining Law of Turkey

Purpose	Article 1
Minerals	Article 2
Definitions	Article 3
Sovereignty and disposition of the State	Article 4
Unity of the rights, their assignment and succession	Article 5
Mining right	Article 6
Permits for mining activities	Article 7
Cases in which mining permits cannot be granted	Article 8, abolished by Law No. 5177
Mining incentive measures	Article 9
Means of declaration	Article 10
Supervision of the activities	Article 11
Production and dispatch	Article 12
Fees, guarantee deposit, fines and other sanctions	Article 13
State rights and local administration's share	Article 14
Discovery right	Article 15
Initial application and licensing	Article 16
Exploration Activity	Article 17
Borders of exploration	Article 18
Pre-operation permit	Article 19, abolished by Law No. 5177
Pre-operation activity	Article 20, abolished by Law No. 5177
Extraction of the ore during the exploration and pre-operation periods	Article 21, abolished by Law No. 5177
Opening of the fields to new explorations	Article 22, abolished by Law No. 5177

Annulment of the pre-operation permit	Article 23, abolished by Law No. 5177
Operation license and operating of the mine	Article 24
Term of operation permit	Article 25, abolished by Law No. 5177
Operation permit	Article 26, abolished by Law No. 5177
Non-assignment of operation permit	Article 27
Starting the operation of the mines	Article 28, abolished by Law No. 5177
Operation activity	Article 29
Tender	Article 30
Technical supervision	Article 31
Nullification of the license and the measures to be taken	Article 32
Transfer of plants	Article 33
Mining fund	Article 34, abolished by Law No. 4629 of 2001
Inspection and supervision expenses	Article 35
Storage of covery, mass of residue and slag	Article 36
Temporary suspension due to force majeure events	Article 37
Organization and particulars of the mine registry	Article 38
Pledge of ores	Article 39
Attachments and precautionary measures	Article 40
Attachments and precautionary measures cannot interfere with the activities of the mine	Article 41
Mortgage and its content	Article 42
Converting mortgage into money	Article 43
Personal liability	Article 44
Rights of easement and usufruct expropriation	Article 46
Rights concerning the services of Mineral Research and Exploration General Directorate	Article 47

Establishment, fields of authority and responsibilities of technical bureaus on oath	Article 48- (Annulled by the decision of Court of Constitution dated: 24/12/1986 with Principal No: 1985/20, Decision No: 1986/30)
	Article 49
Abolished Provisions	Article 51
Mining Department	Article 52 Additional Article 1–5 Provisional Articles 1–12
Effective Date	Article 53

Minerals classification (Article 2 ML)

1st Group of minerals

a) Sand and gravel that are used in general and road construction which are found naturally in nature.

b) Brick-roofing tile clay, cement clay, marl, pozzolanic rocks and rocks that are used in cement and ceramic industries and not included in the other groups.

2nd Group of minerals

Marble, decorative stones, travertine, limestone, dolomite, calcite, granite, cyanide, andesite, basalt and similar stones.

3rd Group of minerals

Salts in the form of solution and obtained from sea, lake and spring water.

4th Group of minerals

a) Kaolin, dickite, nacrite, halloysite, endellit, anauxite, bentonite, montmorillonite, baydilite, nontronite, saponite, hectorite, illite, vermiculite, allofanoid, imalogite, chlorite, sepiolite, palygorskite (attapulgite), loglinite and clays that are a mixture of those, refractory clays, gypsum, anhydrite, potassium, lithium, calcium, magnesium, bromine and other salts, boron salts, strontium salts, barite, vollastonite, talc, pyrophyllite, diatomite, olivine, phosphate, apatite, magnesite, huntite, natural sodium carbonate minerals, zeolite, pumice, stilpnosiderite, perlite, obsidian, grafito, sulfur, fluorite, kryolith, grindstone, corundum, diasporite, quartz, quartzite and quartz sand that contain minimum 80% SiO_2 in its composition, feldspar

b) Peat, lignite, bituminous coal, anthracite, asphaltite, bituminous schist, bituminous shale, radioactive minerals (uranium, thorium, radium)

c) Gold, silver, platinum, copper, lead, zinc, iron, pyrite, manganese, chromium, mercury, vanadium, arsenic, molybdenum, tungsten, cobalt, nickel, cadmium, bismuth, titanium, aluminum (bauxite, gypsite, böhmite), rare earth elements (lanthanides, yttrium) and rare earth minerals bastnaesite, monazite, xenotime, serit, samarskite, fergusonite).

5th Group minerals

Diamond, sapphire, ruby, beryllium, emerald, morganite, aquamarine, and others

Ownership of mineral rights

Minerals are under the sovereignty and disposition of the State, they are not under the proprietorship of the landowners of the land where they exist (Article 4 ML).

Acquiring mineral rights

Mining rights can be granted to the Turkish citizens that are qualified to enjoy civil rights, companies that are legal entities established in accordance with the laws of Turkish Republic and whose statute prescribes that mining is included in their field of activity, public economic enterprises that are authorized on this matter, establishments and administrations (Article 6).

39.3.1.2 Issuing of Permits

Exploration

The permit holder is deemed to be the explorer of the mines declared as proven reserve in the technical reports prepared during the term of exploration and/or operation permits. A discovery certificate will be given to the permit holder requesting this right. In the event the mines that are subject to discovery are operated by another person than the explorer, the explorer rights accrued over the produced minerals on this field has be paid to the right holder by the persons making productions at this filed until the end of June each year. Discovery right is 1% of the annual ex-mine sales price (Article 15).

Minerals of the 2nd, 3rd and 4th Group can be explored by exploration permit and 5th Group minerals can be explored by exploration certificate. For the 1st Group minerals (i. e. construction minerals), operation permit will be granted right away. Applications include a fee (Article 16 ML).

The term of the exploration permit and certificate is 3 years. This term can be extended for the 4th Group minerals for 2 years if the application is made with the exploration activity reports. Permit holder is obliged to submit exploration activity report until the end of the second year. For the 4th Group permits that apply for extension, a second exploration activity report must be submitted at the end of the third year along with the application. In the event exploration activity reports are not submitted in due time, the guarantee deposit must be accounted as revenue (Article 17 ML).

Exploitation

Operation permit right shall be acquired by applying until the end of the exploration permit term with the exploration activity report including the reserve information of the detected minerals, the mining project prepared by the operator and the documentation showing that the application fee is paid. The mining project must ensure the compatibility of the mining area with the environment after the activity (Article 24 ML).

The permit term of 1st Group (a) minerals is minimum five years. Operation permit term of the other group of minerals will be determined according to their projects, provided that this term will not be shorter than ten years. In the event application for extension is made with a new project before the expiry of the term, the permit term can be extended. Total term of the permit cannot exceed 60 years. Council of Ministers is authorized to extend the periods after 60 years (Article 24 ML).

According to Article 29 ML operation activities must be performed in accordance with its project and the related provisions of the Mining Law. It is obligatory to obtain the consent of the General Directorate prior to effectuating the operation projects and amendments thereof. Otherwise, the activities will be suspended. In the event dangerous circumstances are detected with regard to the operation, the permit holder will be granted a six-month period to remedy these circumstances. The permit holder is obliged to submit every year until the end of April of every year the following documents:

- technical documents,
- a sales information form,
- an activity information form regarding the operation activities of the previous year, and

- information regarding exploration if exploration activities were performed at the operation area to the General Directorate.

In the event the obligation is not fulfilled the guarantee deposit will be accounted as revenue. The activity will be suspended until the obligation is fulfilled.

According to Article 30 ML the fields that were abolished, abandoned or reduced for any reason will become available to explorations by putting them out to tender. Announcements of tender will be published in the Official Gazette. In the event there are not any applications during the period of announcement, the area will become available to exploration without having to execute any other transaction. Proceeds of the tender of areas will be accounted as special revenue in the government budget and as special appropriation in the Ministry budget.

Mine registry

A mine registry which will include all the technical and financial issues regarding mining rights and activities must be kept by the General Directorate as specified in the regulation. Mine registry is open to public. Persons concerned may request to view the registry entries at the presence of one of the mine registry officers (Article 38 ML).

39.3.1.3 Authorities

The main responsible authority for mining is the Minister of Energy and Natural Resources (General Directorate of Mines).

According to Article 31 ML production of minerals must be made under the supervision of a mining engineer. The size of operation that requires permanent employment of mining engineer and employment procedures and principles shall be determined by regulations to be issued by the Ministry. Mining engineers employed at the operation must also perform the duties and responsibilities undertaken by engineers or technical staff that are commissioned for industrial safety as specified in Article 82 of Labour Law numbered 4857.

39.3.1.4 Fees and Taxes

It is obligatory to pay fees and guarantee deposit to acquire the permits. Guarantee deposit is 0,3 % of the annual permit fee calculated per hectare depending on the permit stage and permit term. The Council of Ministers is entitled to increase or decrease this rate by 50%. The guarantee deposit amount cannot be less than the annual permit fees determined by the Ministry of Finance each year according to the permit stage and permit term (Article 13 ML).

State rights and local administration's share

The State right to be accrued over produced minerals will be 4% of the ex-mine sales amount for the 1st Group and 5th Group minerals and for any kinds of construction materials such as cover stone, rough construction work, dam, pond, seaport and road. This rate will be 2% for other groups of minerals. Ex-mine sales price declared by the permit the Ministry will inspect holder and incomplete declarations will be corrected. State right will be collected by plus 30% over the mining activities conducted at fields that are in the proprietorship of the Treasury or under the sovereignty and disposition of the State (Article 14 ML). The State rights paid by the permit holder shall be accounted as follows: 50% to the local administration of the city of the permit as the share of local administration, 30% to the account of the Treasury, 20% as special revenue in the government budget and special appropriation to the Ministry budget.

39.3.2 Additional Legal Basics

Table 156: List of Acts significant for raw materials – Turkey

Name of Law	Number and Year of Issuing
Environment Law	Law No. 2872 of 1983
Forest Law	Law No. 6831 as amended by Law Law No. 1987
Law Public Procurement Law	Law No. 4735 of 2002
Law on Work Permits of Foreigner	Law No. 4817 of 2003
Special Provincial Administration Act	Law No. 5302 of 2005
Turkish Civil Code	Law No. 4721 of 2001
Law concerning Arrangements for the implementation of privatization and amending certain laws and decrees with the force of law	Law No. 4046 of 1994
Competition Act	Law No. 4054 of 1994
Customs law	Law No. 4458 of 1999

Environment Law

The objective of the Environment Law is to protect and improve the environment which is the common asset of all citizens; make better use of, and preserve land and natural resources in rural and urban areas; prevent water, land and air pollution; by preserving the country's vegetative and livestock assets and natural and historical richness, organize all arrangements and precautions for improving and securing health, civilization and life conditions of present and future generations in conformity with economical and social development objectives, and based on certain legal and technical principles (Article 1).

40 Ukraine

40.1 General Facts

The national territory covers 603 700 km² and the inhabitants are 46.011.300 (2009 estimation). The population density is 77/km². The GDP per capita amounts to $ 2.542 (2009 estimate).

Constitutional Structure

The form of government is a unitary semi-presidential republic. Ukraine is divided into 24 regions (oblasts). In addition, there are the autonomous region of Crimea, and two cities with special administrative status, Kiev and Sevastopol. The 24 oblasts, the Crimea region are split into 490 districts (rajone). There are 459 towns.

40.2 Production of Raw Materials

Table 157: Production Data – Ukraine (World Mining Data, Weber and Zsak, 2008)

Resources		2002	2003	2004	2005	2006	Change 02/06	Change 05/06
Iron	(t)	35.580.000	37.500.000	39.600.000	44.850.000	43.800.000	23,10	–2,34
Manganese	(t)	1.758.000	1.100.000	940.000	720.000	500.500	–71,53	–30,49
Titanium	(t)	80.000	70.000	66.500	60.000	58.000	–27,50	–3,33
Aluminium	(t)	112.459	113.600	113.200	114.200	113.000	0,48	–1,05
Gold	(kg)	70	70	70	70	70	0,00	0,00
Graphite	(t)	7.431	7.500	7.500	8.000	7.500	0,93	–6,25
Potash	(t)	85.000	75.000	60.000	60.000	65.000	–23,53	8,33
Salt	(t)	2.333.308	2.500.000	2.300.000	2.500.000	2.700.000	15,72	8,00
Sulphur	(t)	91.000	90.000	80.000	78.000	75.000	–17,58	–3,85
Vermiculite	(t)	59.000	60.000	62.000	64.000	65.000	10,17	1,56
Zircon	(t)	35.000	36.000	34.300	35.000	36.000	2,86	2,86

Resources		2002	2003	2004	2005	2006	Change 02/06	Change 05/06
Steam Coal *)	(t)	46.887.000	46.806.000	47.212.000	45.486.500	37.723.000	−19,54	−17,07
Coked Coal	(t)	33.952.000	34.000.000	34.188.000	32.938.500	23.710.000	−30,17	−28,02
Brown Coal	(t)	762.000	700.000	460.000	354.000	278.000	−63,52	−21,47
Natural Gas	(Mm³)	19.000	18.200	18.700	20.100	19.000	0,00	−5,47
Oil	(t)	3.732.500	3.600.000	4.000.000	5.100.000	4.900.000	31,28	−3,92
Uranum	(t)	943	943	943	943	943	0,00	0,00

*) incl. Anthracite

Table 158: Production of clays and gypsum – Ukraine (Metric tons), (USGS, 2008)

	2002	2003	2004	2005	2006
Clays:					
Ball clay	–	–	–	118.000	294.000
Bentonite	300.000	300.000	300.000	300.000	300.000
Kaolin	225.000	225.000	225.000	217.000	251.000
Gypsum	207.000	264.000	337.000	380.600	375.900

40.3 Normative Basics

40.3.1 Primary Legal Basics

The primary legal basics of mineral extraction activity is the Mining Law ("Subsoil Code") No. 132 of 1994. Beside that Law No. 1127 of 1999 is important.

40.3.1.1 General Rules

Table 159: Structure of the Mining Law – Ukraine

Section 1	General provisions (Articles 1–36)
Section 2	Geological studies (Articles 37–41)
Section 3	Registration of deposits (Articles 42–47)

Section 4	Designing, construction and commissioning of mining objects, and also the underground constructions which have been not connected with extraction of minerals (Articles 48–50)
Section 5	Use of subsoil (Articles 51–55)
Section 6	protection of subsoil (Articles 56–59)
Section 7	State control and supervision (Articles 60–63)
Section 8	Dispute settlement and penalties (Articles 64–67)
Section 9	International relations (Articles 68–69)

Table 160: Law No. 1127 of 1999 – Ukraine

Section 1	General Provisions Articles 1–6
Section 2	Mining interests Articles 7–11
Section 3	Preparation for mining and mineral extraction Articles 12–17
Section 4	Exploitation Articles 18–24
Section 5	Avoid damage, protection and safety of mining Articles 25–33
Section 6	Environmental provision Articles 34–36
Section 7	Regulations for working conditions Articles 37–43
Section 8	Termination activities for mining Articles 44–48
Section 9	Infringement regulations Articles 49–51
Section 10	International relations Articles 52–53
Section 11	Final provisions

Ownership of minerals

The state has the responsibility about issuing mineral rights (Article 4 ML). According to the Subsoil Code natural or legal persons from inland and abroad are entitled to acquire mineral rights. Article 9 ML enlarges upon the application areas. All raw materials, which are identified in the state raw material fund, are eligible for licensing. These include iron ore, nonferrous metals, rare earths, and various non-metallic raw materials. Uranium ore, precious metals and gems are listed separately.

The granting of a permit is carried out on the basis of the Law "On Licensing of Certain Types of Economic Activity". The principles of Ukraine's licensing policy are addressed in Article 3 ML.

40.3.1.2 Issuing of Permits

Extraction

Article 12 ML includes provisions for the use of limited resources. For their rational use, technologies should be applied according to the state of the art. The limited resources are incumbent on public tenders. The limited resources include uranium, precious metals and gems as well as gas. The validity of a permit may not fall short of three years and can be renewed in accordance with Article 16 ML. Permits can be transferred to other entities.

Article 16 ML regulates the licensing facts. To obtain a permit the applicant has to be technologically and financially qualified. The ownership must be clarified at the request of a mining permit. The Minister for the Environment grants the exploitation permit in cooperation (with the geological State Committee) through a tender. The Ministerial Decree 1374 of 2004 regulates the pre-definition of fixed charges for tenders.

If a production permit was granted, certain rights and duties are required for the permit holder. The implementation of approved activities and the construction of the necessary facilities will be considered a legal right. The duties include the sustainable and rational use of the subsoil, to ensure the safety of people and the environment.

Article 43 ML directs the register of deposits. This has particular importance for the national raw material balance. This deposit-register must contain information about the quality and the quantity of the deposits. The register is maintained by the State Geological Committee.

Article 44 ML enlarges the economic value of the deposits. This purpose is served by the State Reserve balance; this contains information on quantity and quality of deposits, including industrial value, as well as information on already mined deposits. Article 45 ML determines the methodology for setting deposit ratings. For their identification a special commission is established. Article 52 ML assigns an allocation of individual deposits. For certain types of raw materials the Minister can establish reduction contingents.

Article 56 ML regulates the protection of subsoil. Article 57 ML deals with restrictions and temporary closures of mining activities.

40.3.1.3 Authorities

The main responsible authority for mining is the Minister of Environment. Article 8 ML establishes the regulatory authority in relation to supervision, establishment of tax rates, etc. Article 62–63 ML controls the exact sequence and scope of regulatory supervision. In case of legal violations pursuant to Article 65 ML, generally administrative, civil or penal proceedings will be initiated.

40.3.1.4 Fees and Taxes

Article 28–36 ML regulate fees and charges. Fees are generally due for the use of the subsoil. Permit fees and mining rents are also relevant. The tax rate for exploration activities depends on the kind of raw material, the extent and duration of the activities. The mining rents are set according to the type of raw material and the geological conditions (Article 30 ML). According to Article 35 ML a tax reduction can be granted under certain circumstances. These include exploitation of uneconomic deposits (e.g. minor ore grade) and the use of low-quality remainders.

40.3.2 Additional Legal Basics

Table 161: List of Acts significant for Raw Materials – Ukraine

Name of Law	No. of Law and Year of Issuing
On State Regulation of Mining, Production and use of Precious Metals and Precious Stones and Control over Transactions with them	Law No. 637 of 1997-BP as amended in 2004
Law on Licensing of certain Types of economic Activity	Law No. 1775-III of 2000 as amended 2004
Law on Production Sharing Agreements	1999
Law on Natural Reserves	Law No. 2456-XII of 1992 as amended in 2003
Law on Environmental Protection	Law No. 1264-XII of 1991 as amended in 2003
Law on Land Use Planning and Building	Law No. 1699-II of 2000
Law on Protection of Atmospheric Air	Law No. 2707-XII of 1992 as amended in 2004
Law on Land Protection	Law No. 1877-IV of 2003
Law on Land Reclamation	Law No. 1389-XIV of 2000
Law on the Forms of Land Ownership	Law No. 2073-XII of 1992
Water Code	No. 213/95-VR of 1995 as amended in 2003

Name of Law	No. of Law and Year of Issuing
Law on Waste	Law No. 18 of 1998 as amended in 2002
Administration Act	
Civil Code	Law No. 4721 of 2002, amended by Law No. 4963 of 2003
Law on Companies	

Law on Land Reclamation

Law 1389 of 2000 contains provisions for the land restoration of damaged areas. Separate authorities are responsible for this. The law consists of 33 paragraphs in 11 sections: Section 1 (Articles 1–2) regulates basic provisions, Section 2 (Articles 3–8) relates to the kind of restoration and measures, Section 3 (Articles 9–12) treats property rights, Section 4 (Articles 13–16) discusses the national authorities, Section 5 (Articles 17–18), the local authorities and their executive bodies, Section 6 (Articles 19–23) regulates the implementation of land restoration, Section 7 (Articles 24–27) refers to the conduct and operation of restoration systems, Section 8 (Articles 28–29) regulates the financing, section 9 (Articles 30–31) discusses the ecological compatibility. The final Sections regulate the conflict provisions and final provisions.

41 United Kingdom

41.1 General Facts

The national territory covers 244 820 km² and the inhabitants are 61.113.205 (2009 estimation). The population density is 246/km². The GDP per capita for 2009 is estimated at $ 35.720.

Constitutional Structure

The form of government is a parliamentary democracy and constitutional monarchy. The United Kingdom consists of 4 countries: England, Northern Ireland, Scotland and Wales and has 14 overseas territories. It has a high level of development.

41.2 Production of Raw Materials

Table 162: Production Data – UK (World Mining Data, Weber and Zsak, 2008)

Resources		2002	2003	2004	2005	2006	Change 02/06	Change 05/06
Iron	(t)	255	275	280	500	550	115,69	10
Aluminium	(t)	344.318	342.748	359.600	368.477	360.300	4,64	–2,22
Cadmium	(t)	292	22	0	0	0	–100,00	
Lead	(t)	700	700	500	500	500	–28,57	0,00
Baryte	(t)	59.000	57.000	57.000	62.000	44.000	–25,42	–29,03
Bentonite	(t)	44.000	34.000	30.000	28.000	20.000	–54,55	–28,57
Diatomite	(t)	3.000	4.000	4.300	4.400	3.000	0,00	–31,82
Feldspar	(t)	3.000	2.000	1.800	2.000	2.500	–16,67	25,00
Fluorspar	(t)	53.000	56.000	56.000	60.980	60.000	13,21	–1,61
Gypsum	(t)	1.700.000	1.700.000	1.750.000	1.700.000	1.700.000	0,00	0,00
Kaolin	(t)	2.162.815	2.094.137	2.097.000	1.908.000	1.762.000	–18,53	–7,65
Potash	(t)	540.100	621.400	650.000	439.200	716.000	32,57	63,02
Salt	(t)	6.100.000	5.900.000	5.700.000	5.800.000	5.224.000	–14,36	–9,93
Sulfur	(t)	125.000	115.000	110.000	124.000	126.000	0,80	1,61

Resources		2002	2003	2004	2005	2006	Change 02/06	Change 05/06
Talc	(t)	5.000	6.000	6.300	6.000	4.000	−20,00	−33,33
Steam-Coal	(t)	29.616.000	27.886.000	24.606.000	20.350.000	17.501.000	−40,91	−14,00
Coking-Coal	(t)	373.000	373.000	371.000	274.000	266.000	−28,69	−2,92
Nat. Gas	(Mm³)	109.512	108.700	100.500	92.800	83.400	−23,84	−10,13
Oil	(t)	107.400.000	97.820.000	92.000.000	84.500.000	75.200.000	−29,98	−11,01

Remarks: Steam coal incl. anthracite

Table 163: Aggregates Production – Annual Statistics/United Kingdom, 22 April 2008, quantities in million tonnes (UEPG, 2008)

Sand & Gravel (1)		Crushed rock (2)		Marine Aggregates		Recycled Aggregates (3)		Manufactured Aggregates (4)	
2004	2006	2004	2006	2005	2006	2005	2006	2005	2006
124,0	68,0	85,0	123,0	n. a.	13,0	56,0	58,0	12,0	12,0

(1) Sand and Gravel: sold production including crushed gravel
(2) Crushed rock: sold production (excluding crushed gravel)
(3) Recycled Aggregates: materials coming from construction and demolition waste used in aggregates market
(4) Manufactured aggregates include blast-furnace-slag, electric-arc-furnace-slag, incinerator bottom ash (IBA), pulverised fuel ash (PFA)

41.3 Normative Basics

41.3.1 Primary Legal Basics

There are different laws relevant as legal basics of mineral extraction activity.

41.3.1.1 General Rules

Classification of minerals

In the UK "minerals" are defined in Town and Country Planning legislation as: *all substances in or under land of a kind ordinarily worked for removal by underground or surface working, except that it does not include peat cut for purposes other than for sale* (http://www.bgs.ac.uk/mineralsuk/planning/legislation/uk_minown.html).

Ownership of minerals

There is no general state ownership of mineral rights in the UK, except for energy minerals (oil, gas and coal) and precious metals (gold and silver).

The rights to gold and silver in most of the UK are owned by the Crown. Mines of these metals are known as "Mines Royal". A licence, known as a Mines Royal Permit, for the exploration and development of these metals must be obtained from the Crown Estate Commissioners through the Crown Mineral Agent. The rights to gold and silver in the former county of Sutherland in northern Scotland are held by the Duchy of Sutherland (http://www.bgs.ac.uk/mineralsuk/planning/legislation/uk_minown.html).

There is no national register of mineral ownership, but the Land Registry may have details of surface ownership and current ownership of mineral rights. There is also no national licensing system for exploring for and extracting these minerals. However, all mineral development and, in some cases, exploration requires planning permit from a mineral planning authority.

41.3.1.2 Issuing of Permits[1]

Exploration

Part 22 of Schedule 2 of the GPDO (General Permitted Development Order) 1995 allows certain small scale and temporary mineral exploration activities to be undertaken without planning permission. Those activities are: Drilling of boreholes (applies to small scale operations not to drilling for, for example, oil and gas); the making of other excavations; the carrying out of seismic surveys (Department of the Environment, Transport and Regions 1996). These activities are permitted for up to 28 days.

Class B of part 22 permits the same operations to be carried out for the longer period of 6 months. This permit is indefinitely renewable in respect of the same land by the service of further notice of the Mineral Planning Authority. Through this procedure flexibility is provided to the mineral operators, which is needed to adjust their exploration programme to the size and complexity of the prospect. The exploration permit may only be exercised in accordance with the details specified in the written notice given to the MPA (Department of the Environment, Transport and Regions, 1996).

The Town and Country Planning Order of 1992 applies similar arrangements in Scotland.

---

[1] All information from Ike, P. (2004), Country Report Great Britain, in: Department of Mining and Tunneling, lc.

Exploitation

The planning system in the UK has greatest effect at the exploitation stage. In England and Wales there is a presumption in favour of permission being granted for development that is in accord with the Development Plan (see clause 38(6) of the 2004 Planning and Compulsory Purchase Act). Development plans for minerals are produced by each mineral planning authority. In England this could be a county authority or, as in Wales and Scotland, a unitary planning authority. The planning system in Northern Ireland is administered centrally. Each development plan aims to make provision for an adequate and steady supply of minerals.

Planning permission for mineral extraction in England, Wales and Scotland is granted by the mineral planning authority.

The application is publicly advertised with site notices and through the press. Before the MPA reaches a decision on an application, it consults interested parties and takes into account their views. (Department of the Environment, Transport and Regions, 1996)

An operator is enabled to enter into a planning agreement with a local planning authority (legal agreements). He can require specified operations to be carried out in, on, under land; require the land to be used in a specific way; require a sum or sums to be paid to the authority on a specified date or dates periodically.

All permits must have a time limit condition, requiring development to cease not later than the expiration of 60 years or such longer or shorter period as the Mineral Planning Authority may specify. Permits existing on 22 February 1982, which are not already time-limited, become time-expired on 22 February 2042.

The overarching minerals policy and mineral planning policy for England is set out in Minerals Policy Statement 1: Planning and Minerals (CLG 2006). MPS1, together with other policy statements and guidance provided a context for decision making on development plan provision for new working areas and on the determination of individual applications.

Mineral Planning Guidance 2 ("Applications, permissions and conditions") provides detailed guidance on planning conditions. These are: Time Limits, access and protection of the public highway, extraction plans, buildings, fixed plant and machinery, environmental protection, surface water, drainage and pollution control, landscaping, boundaries and site security, restoration and aftercare, subsidence and support (Department of the Environment 1995). Most of MPG2 (1988) still applies in Wales.

Mineral Planning Guidance 8 states that applications for registration of the permit must be made on an official form obtainable from the MPA and must be accompanied by the appropriate certificates that the necessary persons have been properly notified of the application, or that the application has been properly advertised.

Scotland

Extraction permit is required for all mining operations in, on, or under land. Application should be accompanied by: a plan sufficient to identify the land concerned; other plans, drawings and additional documentation to describe the development proposal; environmental statement (ES) if required; proposals for restoration and aftercare of a site; and appropriate certificates and fees.

When deciding whether to grant permit Mineral Planning Authorities need to have regard to the provisions in the development plan for the area. Material considerations will vary on a case by case basis.

If approved, Mineral Planning Authorities should attach appropriate planning conditions to regulate the development. A sensitive use of appropriate planning conditions can provide important environmental safeguards sufficient to mitigate any environmental effects through sound working practices and restoration and aftercare procedures. The Town and Country Planning (Scotland) Act 1997 includes legislative provisions for planning authorities to review regularly the conditions attached to all mineral permit so that improved operating and environmental standards can be secured.

Restoration

The statutory definitions of restoration and aftercare in England are established in Schedule 5 of the 1990 Town and Country Planning Act. MPS1 includes as a national objective for minerals planning, "to protect and seek to enhance the overall quality of the environment once extraction has ceased, through high standards of restoration, and to safeguard the long-term potential of land for a wide range of after-uses". National mineral planning policy 19 in that document provides a more detailed context for the consideration of mineral restoration issues. MPG7: The Reclamation of Mineral Workings gives further guidance on the subject. The Mineral Planning Authorities (MPAs) draw up the restoration and aftercare conditions and are the responsible authority in monitoring the mineral operators activities. Careful lifting, storage and replacement of soil in the appropriate order and to appropriate depths can be required by the restoration conditions (Department of the Environment, Transport and Regions, November 1996). Schedule 5 of the 1990 Town and Country Planning Act provides powers to the MPAs to impose aftercare conditions on the grant of planning permits in relation to land, which is to be used for: Agriculture, forestry, amenity (Department of the Environment, Transport and Regions, November 1996).

The statutory basis is the same in Wales as for England. MPPW (December 2000) states that unless new mineral extraction provides satisfactory and suitable restoration permit should be refused. Planning conditions should ensure that sites are restored to a high standard within 6 months of cessation of working wherever practicable and provide the means to maintain and preferably enhance the long-term quality of land used for mineral extraction.

41.3.1.3 Authorities

In England, the main responsibility for mineral planning policy rests with the Department for Communities and Local Government, in Wales the Welsh Assembly Government and in Scotland the Scottish Parliament.

Inspections on compliance with extracting minerals, restoration and aftercare conditions will be made by the MPAs as a basis for enforcement action if necessary.

Most companies in England and Wales have some form of environmental policy or code. To the same end, most of these companies subscribe to a code from their trade association. In most of the cases these codes are not formal. The Government welcomes codes of practice drawn up by the minerals industry and encourages mineral companies to adopt environmental management systems (Department of the Environment, Transport and Regions 1996b).

Mineral Planning authorities in Scotland are responsible for monitoring the conditions that are attached to planning permit. The enforcement powers available to planning authorities for mineral working are generally similar to all other forms of development. The main provisions are set out in Part VI of the Town and Country Planning (Scotland) Act 1997.

The Mineral Planning Authority must make sure that the conditions attached to a permission are being adhered to. The Mineral Planning Authorities have several powers to enforce planning permits: Revocation or modification of a permit with regard to the development plan and to other mineral considerations. The planning authority may prohibit the resumption of mineral extraction and impose certain requirements including removal of plant and restoration.

41.3.1.4 Fees and Taxes

In Great Britain an Aggregates Levy was implemented. The objective of this levy is to help to offset the environmental costs associated with aggregates quarrying operations in line with the Government's statement of intent on environmental taxation. The Levy is intended to reduce waste of construction materials and encourage the use of re-cycled materials, thus reducing demand for virgin aggregate. The scope and structure of this levy is that it applies to quarried and dredged sand, gravel and crushed rock subjected to commercial exploitation in the UK.

Imported apprepates are also subject to the Aggregates Levy when they are landed.

41.3.2 Additional Legal Basics

Table 164: List of Acts significant for raw materials – UK (Ike, 2004)

Name of Law	Year of Issuing
UK:	
England: (Department of the Environment, Transport and Regions)	1996
The Commons Act	1876
Metropolitan Commons Act	1866–1898
Law of Property Act	1925
National parks and Access to the Countryside Act	1949
Mineral Workings Act	1951
Mines And Quarries Act	1954
Mines (Working Facilities and Support) Acts	1966 and 1974
Mines and Quarries (Tips) Act	1969
Mineral Workings Act	1971
Local Government Act	1972;
Health and Safety at Work Act	1974
Local Government (Miscellaneous Provision) Act	1976
Ancient Monuments and Archaeological Areas Act	1979

41.3 Normative Basics

Name of Law	Year of Issuing
Local Government, Planning and Land Act	1980
Wildlife and Countryside Act	1981
Road Traffic Regulation Act	1984
Mineral Workings Act	1985
Planning (listed Buildings and Conservation Areas) Act	1990
Environmental Protection Act	1990
Planning (Hazardous Substances) Act	1990
Town and Country Planning Act	1990
Planning and Compensation Act	1991
Water Resources Act	1991
Land Drainage Act	1991
Coal Mining (Subsidence) Act	1991
Noise and Statutory Nuisance act	1993
Railways Act	1993
Coal Industry Act	1994
Environment Act	1995
Aggregates Levy [this is a tax NOT legislation]	2002
Planning and Compulsory Purchase Act	2004
Wales: The laws listed in the section "England" are also applicable to Wales.	
Scotland:	
Principal Planning legislation: Town and Country Planning (Scotland) Act	1997
Town and Country Planning (General Development Procedure) (Scotland) Order	1992
Town and Country Planning (General Permitted Development) (Scotland) Order	1992
Town and Country Planning (Structure and Local Plans) (Scotland) Regulations	1983
Other main relevant laws and regulations, Environmental Assessment (Scotland) Regulations	1999
Mines (Working Facilities and Support) Acts	1966 and 1974
Local Government (Scotland) Act	1973
Local Government etc. (Scotland) Act	1994
Local Government in Scotland Act	2003
Health and Safety at Work Act	1974

Name of Law	Year of Issuing
Wildlife and Countryside Act	1981
Natural Heritage (Scotland) Act	1991
Nature Conservation (Scotland) Act	2004
Environmental Protection Act	1990
Coal Industry Act	1994
Environment Act	1995
Water Industry (Scotland) Act	2002
Water Environment and Water Services (Scotland) Act	2003

42 References

A

Agioutantis, Z. (2004), Country Report Greece, in: Department of Mining and Tunnelling, University of Leoben

B

BGR (Bundesanstalt für Geowissenschaften und Rohstoffe; Federal Institute for Geosciences and Natural Resources), 2009. Hannover, Germany. http//www. bgr. bund. de

Badino, G. et al (2004), Country Report Italy, in: Department of Mining and Tunnelling, University of Leoben

Bauer, V. (2004), Country Report Czech Republic, in: Department of Mining and Tunnelling, University of Leoben

Bauer, V. (2004), Country Report Slovakia, in: Department of Mining and Tunnelling, University of Leoben

C

Chodak, M. (2004), Country Report Latvia, in: Department of Mining and Tunnelling, University of Leoben

D

Department of the Environment (1995): Minerals Planning Policy and Supply Practices in Europe, Technical Appendices, London.

Department of Mining and Tunnelling, University of Leoben (2004): Minerals Policies and Supply Practices in Europe, Commissioned by the European Commission Enterprise Directorate General under Contract n° ETD/FIF 2 003 0781, Leoben – Brüssel, November 2004

E

Espi, J. (2004), Country Report Spain, in: Department of Mining and Tunnelling, University of Leoben

European Commission (DG Enterprise), (2007). Analyses of the competitiveness of the non-energy industry. Commission staff working document, Brussels.

H

Hamor, T. (2004): Country Report Hungary, in: Department of Mining and Tunnelling, University of Leoben.

I

Ike, (2004): Country Report Irland, in: Department of Mining and Tunnelling, University of Leoben.

Ike, (2004): Country Report Great Britain, in: Department of Mining and Tunnelling, University of Leoben.

Ike, (2004): Country Report Netherlands, in: Department of Mining and Tunnelling, University of Leoben.

K

Koziol, W., Kawalec, P. (2004), Country Report Estonia, in: Department of Mining and Tunnelling, University of Leoben

Kullmann, U., Requirements for a modern mining law in the framework of European legislation, Federal Ministry of Economics and Technology, Germany

Kündig et al, (1997), Die mineralischen Rohstoffe der Schweiz, Zürich

M

Magna, C, (2004), Country Report Portugal, in: Department of Mining and Tunnelling, University of Leoben

Maier, A. (2005), Reflexionen zur österreichischen Entwicklung von Bergrecht und Bergbehörden, Berg- und Hüttenmännische Monatshefte, Jg. 150, S. 438–448.

Mihatsch, A. (2002), Mineralrohstoffgesetz (MinroG). 2., überarbeitete Auflage. MANZ'sche Verlags- und Universitätsbuchhandlung GmbH, Wien.

Müller, W und Schulz, P.-M. (2000) Handbuch Recht der Bodenschätzegewinnung, Baden-Baden, Deutschland.

N

Nielsen, K (2004), Country Report Denmark, in: Department of Mining and Tunnelling, University of Leoben.

Nielsen, K (2004), Country Report Finland, in: Department of Mining and Tunnelling, University of Leoben.

Nielsen, K (2004), Country Report Norwegian, in: Department of Mining and Tunnelling, University of Leoben.

Nielsen, K (2004), Country Report Sweden, in: Department of Mining and Tunnelling, University of Leoben.

O

Otto, J. M. (1999): Mining, Environment and Development, United Nations Conference on Trade and Development, USA

S

Solar, S. (2004): Country Report Slovenia, in: Department of Mining and Tunnelling, University of Leoben

T

Tiess, G. (2004), Country Report Austria, in: Department of Mining and Tunnelling, University of Leoben.

Tiess, G. (2004), Country Report Switzerland, in: Department of Mining and Tunnelling, University of Leoben

Tiess, G. (2010): General and International Mineral Policy – Focus: Europe, Springer-Verlag, Wien

U

Ubermann, R., Ostrega, A. (2004), Country Report Lithuania, in: Department of Mining and Tunnelling, University of Leoben

Ubermann, R., Ostrega, A. (2004), Country Report Poland, in: Department of Mining and Tunnelling, University of Leoben

UEPG Union Européenne des Producteurs de Granulats (2008), Baurohstoffstatistiken, http://www.uepg.eu/uploads/documents/122–51 uepg_statistics_2005-de.xls, http://www.uepg.eu/uploads/documents/141–99-statistics2006de.xls

USGS (Unites States Geological Survey) (2008): Minerals Yearbook.

V

Vervoort, A. (2004), Country Report Belgium, in: Department of Mining and Tunnelling, University of Leoben

Vervoort, A. (2004), Country Report France, in: Department of Mining and Tunnelling, University of Leoben

Vervoort, A. (2004), Country Report Luxembourg, in: Department of Mining and Tunnelling, University of Leoben

Vereinigung Rohstoffe und Bergbau (VRB) (2008), 2008, Positionen und Perspektiven.

W

Weber, L., Zsak, G. (2008): World Mining Data 2006, Bundesministerium für Wirtschaft und Arbeit, Wien

43 Additional updated data directly from the concerned countries

Czech Republic

Outline of domestic mine production

		2005	2006	2007	2008	2009
Energy minerals						
Uranium	t U	420	383	322	290	286
	Concentrate production, t U*	409	358	291	261	243
Bituminous coal	kt	12.778	13.017	12.462	12.197	10.621
Brown coal	kt**	48.658	48.915	49.134	47.456	45.354
Lignite	kt	467	459	437	416	262
Crude oil	kt	306	259	240	236	217
Natural gas	mil m³	356	148	148	168	180
Industrial minerals						
Graphite	kt	3	5	3	3	0
Pyrope bearing rock	kt	43	39	34	24	26
Moldavite (tectite) bearing rock	ths m3	74	95	114	99	58
	kt (1 m³ = 1.8 kt)	133	171	205	177	104
Kaolin	Raw, kt ***	3.882	3.768	3.604	3.833	2.886
	Beneficiated, kt	649	673	682	672	525
Clays	kt	671	561	679	574	377
Bentonite****	kt	216	267	335	235	181
Diatomite	kt	38	53	19	31	0
Feldspar	kt	472	487	514	488	431
Feldspar substitutes	kt	23	31	25	36	23
Silica minerals	kt	15	17	19	18	16

		2005	2006	2007	2008	2009
Glass sand	kt	920	963	942	1.151	990
Foundry sand	kt	807	773	850	702	374
Limestones and corrective additives for cement production	kt	10.190	10.441	11.670	11.465	9.488
Dolomite	kt	419	409	385	449	337
Gypsum	kt	25	16	66	35	13
Construction minerals						
Dimension stone	Mine production in reserved deposits, ths m3 *****	288	242	242	229	209
	Mine production in reserved deposits, kt (1 m³ = 2.7 kt)*****	778	653	653	618	564
	Mine production in non-reserved deposits, ths m³ ******	55	55	50	45	54
	Mine production in reserved deposits, kt (1 m³ = 2.7 kt)******	149	149	130	105	146
Crushed stone	Mine production in reserved deposits, ths m³ *****	12.822	14.093	14.655	14.799	13.947
	Mine production in reserved deposits, kt (1 m³ = 2.7 kt)*****	34.619	38.051	39.569	39.957	37.657
	Mine production in non-reserved deposits, ths m³ ******	1.270	1.300	1.350	1.600	1.350
	Mine production in non-reserved deposits, kt (1 m³ = 2.7 kt)******	3.429	3.510	3.645	4.320	3.650

43 Additional updated data directly from the concerned countries

		2005	2006	2007	2008	2009
Sand and gravel	Mine production in reserved deposits, ths m³ *****	9.075	9.110	9.185	8.770	7.269
	Mine production in reserved deposits, kt (1 m³ = 1.8 kt)*****	16.335	16.398	16.533	15.786	13.084
	Mine production in non-reserved deposits, ths m³ ******	5.100	6.000	6.450	6.350	6.050
	Mine production in non-reserved deposits, kt (1 m³ = 1.8 kt)******	9.000	10.800	11.700	11.520	10.890
Brick clays and related minerals	Mine production in reserved deposits, ths m³ *****	1.543	1.286	1.433	1.242	1.028
	Mine production in reserved deposits, kt (1 m³ = 1.8 kt)*****	2.777	2.315	2.579	2.236	1.850
	Mine production in non-reserved deposits, ths m3 ******	220	290	300	270	203
	Mine production in non-reserved deposits, kt (1 m³ = 1.8 kt)******	396	540	540	520	365
Metallic ores (not mined)						

* corresponds to sales production (without beneficiation losses)
** ČSÚ (Czech Statistical Office) presents so-called sales mining production which is production of marketable brown coal and reaches on average about 95 % of given mine production
*** raw kaolin, total production of all technological grades
**** including mining of montmorillonite clays overburden of kaolins since 2004
***** decrease of mineral reserves by mining production
****** estimate

Spain

Spain Mining Production

Spain		2004	2005	2006	2007	2008	Change 08/07, %
Iron	t (metal content)	7 000	0	0	0	0	–
Cooper	t (metal content)	0	4.840	6.480	6.510	7.070	8,6
Niquel	t (metal content)	0	5.320	6.340	6.770	8.130	20,0
Tungsten	t (metal content)	0	0	0	0	194	–
Gold	kg (metal content)	5.248	4.016	3.403	0	0	–
Silver	kg (metal content)	3.583	4.416	3.104	0	0	–
Steam-Coal	10^3 t	11.338	11.887	11.569	11.000	10.129	–7,9
Lignite	10^3 t	8.147	7.587	6.860	6.209	0	–
Natural Gas	106 Nm³	370	174	118	42	46	14,3
Oil	10^3 t	263	174	143	143	128	–10,5
Refractory clays	10^3 t (Al_2O_3)	36	55	84	97	88	–9,2
Attapulgite	10^3 t	21	21	21	25	27	8,0
Barite	10^3 t ($BaSO_4$)	34	35	37	22	10	–54,5
Bentonite	10^3 t	157	163	155	147	139	–5,0
Washed Kaolin	10^3 t (Al_2O_3)	145	156	151	159	124	–22,0
Quartz	10^3 t (SiO_2)	1.123	1.213	1.273	1.200	884	–26,3
Fluorspar	10^3 t (F2Ca)	132	134	137	140	132	–5,7
Talc	10^3 t	108	91	84	78	59	–24,3
Celestine	10^3 t ($SrSO_4$)	193	312	263	128	125	–2,3
Feldspar	10^3 t	553	650	675	683	690	1,0
Sodium sulphate	10^3 t (Na_2SO_4)	1.106	1.110	1.052	1.096	1.104	0,7
Magnesite	10^3 t (MgO)	241	237	222	197	188	–4,5
Mica	10^3 t	8	8	13	16	14	–12,5
Iron pigments	10^3 t (Fe_2O_3)	89	103	50	65	60	–7,6
Salt	10^3 t	3.994	4.400	4.406	4.145	4.303	3,8
Potash	10^3 t (K_2O)	553	495	493	532	473	–11,1
Sepiolite	10^3 t	655	808	806	718	738	2,7
Tripoli	10^3 t (SiO_2)	29	40	40	40	38	–5,0

Spain		2004	2005	2006	2007	2008	Change 08/07, %
Peat	10^3 t	57	73	88	87	81	−6,8
Gypsum	10^3 t	12.534	14.453	15.330	14.536	13.300	−8,5
Dunite	10^3 t	1.093	970	1.775	1.379	1.500	8,7
Ornamental Granite	10^3 t	1.458	1.576	1.762	1.915	1.142	−40,3
Marble	10^3 t	4897	4.693	4.911	4.677	4.316	−7,7
Roof Slate	10^3 t	839	907	916	935	625	−33,1

Sweden

Aggregates production 2006 – Annual Statistic Sweden, November 2007 (quantities in million tonnes), (Production and resources 2006 SGU Per.publ. 2007:3, Geological Survey of Sweden)

Sand & Gravel		Crushed rock		Marine Aggregates		Recycled Aggregates		Manufactured Aggregates	
2005	2006	2005	2006	2005	2006	2005	2006	2005	2006
19,9	19,9	62,3	71,3	0	0	n.a.	n.a.	n.a	n.a.

Production Date-Sweden (Source: Statistic of the Swedish Mining Industry 2009 October 2010 SGU Per.publ. 2010:1)[1], (Production and resources 2009 SGU Per.publ. 2010:2)[2], Geological Survey of Sweden, World Bureau of Metal Statistics[3]

		2002	2003	2004	2005	2006	2007	2008	2009
Iron[1]	(Mt)	20,3	21,5	22,3	23,3	23,3	24,7	23,9	17,7
Aluminium[3]	(Kt)	100,6	101,2	101,4	102,5	101,2	100,1	112,0	70,0
Copper[1]	(Kt)	72,0	83,1	82,4	87,1	86,7	62,9	57,7	55,4
Lead[1]	(Kt)	43,0	51,0	54,3	60,4	55,6	63,2	63,5	69,3
Zinc[1]	(Kt)	148,6	185,6	197,0	215,7	210,0	214,6	188,0	192,5
Gold[1]	(t)	5,8	5,6	6,6	6,6	6,8	5,2	4,9	5,5
Silver[1]	(t)	320,8	340,7	319,6	309,9	292,3	323,2	293,1	288,6
Feldspar[2]	(Kt)	37	44	38	30	24	25	22	18
Talc[2]	(Kt)	20	7	8	7	6	7	4	4